Daniel S. Alexander

A History of Complex Dynamics

Aspects of Mathematics

Edited by Klas Diederich

*A Publication of the Max-Planck-Institut für Mathematik, Bonn

Volumes of the German-language subseries "Aspekte der Mathematik" are listed at the end of the book.

Daniel S. Alexander

A History of Complex Dynamics

From Schröder to Fatou and Julia

vieweg

Professor Daniel S. Alexander
Department of Mathematics
Drake University
203 Howard Hall
Des Moines, Iowa 50311-4505
USA
da 0231 r © acad. drake. edv

Mathematics Subject Classification: 01–02, 01 A 60, 01 A 55, 30 D 05, 30 C XX

Originally published by Friedr. Vieweg & Sohn Verlagsgesellschaft mbH, Braunschweig/Wiesbaden in 1994
Softcover reprint of the hardcover 1st edition 1994

Cover design: Wolfgang Nieger, Wiesbaden
Printed on acid-free paper

ISSN 0179-2156

ISBN 978-3-663-09199-8 ISBN 978-3-663-09197-4 (eBook)
DOI 10.1007/978-3-663-09197-4

Contents

List of Figures

Preface

In late 1917 Pierre Fatou and Gaston Julia each announced several results regarding the iteration of rational functions of a single complex variable in the *Comptes rendus* of the French Academy of Sciences. These brief notes were the tip of an iceberg. In 1918 Julia published a long and fascinating treatise on the subject, which was followed in 1919 by an equally remarkable study, the first installment of a three-part memoir by Fatou. Together these works form the bedrock of the contemporary study of complex dynamics.

This book had its genesis in a question put to me by Paul Blanchard. Why did Fatou and Julia decide to study iteration? As it turns out there is a very simple answer. In 1915 the French Academy of Sciences announced that it would award its 1918 *Grand Prix des Sciences mathématiques* for the study of iteration. However, like many simple answers, this one doesn't get at the whole truth, and, in fact, leaves us with another equally interesting question. Why did the Academy offer such a prize?

This study attempts to answer that last question, and the answer I found was not the obvious one that came to mind, namely, that the Academy's interest in iteration was prompted by Henri Poincaré's use of iteration in his studies of celestial mechanics. While this may have played a part in the Academy's decision, it also turns out that there was a longstanding French interest in the iteration of complex maps, beginning with the studies of Gabriel Koenigs in the mid-1880's. However, he was not the first to become intrigued by the dynamics of complex maps. That honor seems to belong to a German mathematician, not unknown by any means, but one who deserves more renown than he seems to have at the present moment, Ernst Schröder, who in 1870 articulated the following theorem.

> Let $\phi^n(z)$ denote the n-fold composition of $\phi(z)$ with itself. If $\phi(z)$ is a complex analytic map satisfying $\phi(x)$ and $|\phi'(x)| < 1$ for some point x, then there exists a neighborhood D of x on which $\phi^n(z)$ converges to x for all z in D.

This book traces the history of the iteration of complex maps from Schröder's first paper to the studies of Fatou and Julia. I have tried to keep myself focused on that development, and as a consequence decided not to include a number of interesting topics that I judged off that beam. Schröder's work in particular is worthy of a more detailed treatment than I have given it here.

The one area where I did let myself wander a bit was with regard to the development of Paul Montel's theory of normal families, without which the studies of Fatou and Julia would have been significantly different. In fact, a glimpse of what complex dynamics might look like without this theory is offered in the work of Samuel Lattès and Joseph Fels Ritt, each of whom also investigated the iteration of complex maps around the end of World War I, but without the benefit of Montel's theory. The theory of normal families is such an important component of the studies of Fatou and Julia that I indulged myself and included a brief outline of its development since I had found so little information on it elsewhere. In order to stay on track, however, I did not say as much about Montel as I might have liked, a situation I hope to soon remedy.

Before getting underway, I would like to thank several people who have provided help and inspiration. First of all, I want to thank my advisor, Tom Hawkins, whose encouragement and advice have been an immeasurable benefit. I am also greatly indebted to those at Boston University who helped me develop an interest in this most beautiful subject, especially Paul Blanchard, Bob Devaney and Dick Hall. John Erik Fornæss and Tom Scavo, reader extraordinaire, have made many valuable comments regarding my manuscript during its various stages. My deep gratitude is also extended to Brigette Döbert and Klas Diederich at Vieweg, to Pierre Lelong and Pierre Dugac for their help in obtaining archival information from the French Academy of Sciences, as well as to the mathematics departments of Boston University, Colby College and Drake University for their generous support during the preparation of my manuscript. Nor do I wish to forget Jerry, Bobby and Phil for the tunes which have often provided a much needed escape.

I have saved the best for last: my deepest thanks go to Rebecca and Caroline for tolerating my frequent absences, and to my wonderful parents, to whom I dedicate this work.

Chapter 1

Schröder, Cayley and Newton's Method

1.1 Introduction

The body of work on the iteration of complex analytic functions which culminated in the major studies of Fatou and Julia has its origins in two detailed examinations of Newton's method. The first was a remarkable paper by the German mathematician Ernst Schröder (1841–1902), published in two parts in 1870 and 1871, and the second, written by the British mathematician Arthur Cayley (1821–1895), appeared in 1879.

Newton's method, probably the oldest and most famous iterative process to be found in mathematics, can be used to approximate both real and complex solutions to the equation $f(z) = 0$. Picking an initial value z_0 near a root of the equation, Newton's method produces an *nth* approximation of the root via the formula

$$z_{n+1} = z_n - \frac{f(z_n)}{f'(z_n)}. \tag{1.1}$$

Replacing z_n by z_{n+1} generates a sequence of approximations $\{z_n\}$ which may or may not converge to a root of the equation $f(z) = 0$.

Versions of Newton's method had been in existence for centuries previous to the studies of Cayley and Schröder. Anticipations of Newton's method are found in an ancient Babylonian iterative method of approximating the square root of a,

$$z_{n+1} = \frac{1}{2}(z_n + \frac{a}{z_n}),$$

which is equivalent to Newton's method for the function $f(z) = z^2 - a$, as well as in François Vièta's (1540–1603) use of an iterative method equivalent to Horner's method.

Despite its name, Issac Newton (1642–1727) did not present the algorithm known as Newton's method in the form given at equation (1.1). Indeed, Newton himself may be said to have anticipated equation (1.1), since his explicit formulation was not the one usually associated with Newton's method. Both Ypma in his paper [1993] and Kellerstrom in [1992] observe that although the procedure Newton gave was equivalent to (1.1), it was not formulated in terms of fluxions but instead given algebraically.

Joseph Raphson (1648–1715) also described a method equivalent to (1.1), which he claimed he developed independently of Newton. Like Newton, however, his formulation did not explicitly involve the calculus. The first to bring the derivative into the picture was Thomas Simpson (1710–1761). Ypma gives sound reasons for calling (1.1) the Newton-Raphson-Simpson method, but as Kellerstrom points out, Joseph Fourier (1768–1830) evidently was the first to express this method explicitly as (1.1), and some therefore refer to it as the Newton-Fourier method. For the sake of brevity, I will refer to it simply as Newton's method.

All told, by the mid-1800's several mathematicians had already examined the convergence of Newton's method towards the real roots of an equation $f(z) = 0$, but the investigations of Cayley and Schröder are distinguished from their predecessors in their consideration of the convergence of Newton's method to the complex roots of $f(z) = 0$.

Schröder and Cayley each studied the convergence of Newton's method for the complex quadratic function, and both showed that on either side of the perpendicular bisector of the roots, Newton's method converges to the root on that particular side. Schröder established this theorem in the second of his two papers on iteration, the paper [1871]. Despite this fact, Cayley is often credited with the first proof of this result even though his proof—which he evidently accomplished without knowledge of Schröder's work—did not appear until 1879.

The scope of Schröder's study was quite a bit wider than was Cayley's work. Schröder's treatment of Newton's method was one part of a general discussion of iterative equation solving algorithms, which included an investigation of the process of iteration itself wherein he discovered several fundamental concepts regarding the iteration of complex functions. Cayley's aims were far more modest, and he confined himself entirely to the study of Newton's method.

1.2 Schröder's Study of Iteration

Ernst Schröder received his doctorate in mathematics from the University of Heidelberg in 1862. Among those he studied under was Ludwig Otto Hesse (1811-74). He studied both mathematics and physics at the University of Königsberg for the next two years and subsequently taught at various secondary schools. Beginning in 1874 he began teaching at the college level, first at the Technical Institute of Darmstadt and then in 1876 at the Technical Institute of Karlsruhe, where he evidently remained for the duration of his career.

Schröder's principal fields of research were logic and set theory, and he was an early proponent of the works of Georg Cantor (1845–1918). However, early in his mathematical career he became quite interested in the iteration of complex functions and published two important papers on the subject, [1870] and [1871]. These two works comprise his entire research output on the iteration of complex functions.

Although he was speaking primarily of Schröder's research in logic and set theory, the historian Wussing observed a certain "prolixity" [1980:216] in Schröder's work, a word which can well be applied to portions of his papers [1870] and [1871]. Moreover, Wussing noted that Schröder's work in logic and set theory, although "in the mainstream of the conceptual development of mathematical logic" did not meet with immediate acceptance, due in part to difficulties in his style and in part to the fact that he spent most of his life away from the centers of German mathematical research teaching in technical colleges [1980:216].

Nonetheless, his *Vorlesungen übe die Algebra der Logik*, a multi-volume work published between 1890 and 1905, came to be considered a classic and remains in print to this day. Another indication of his accomplishments in set theory is the fact that an important set theoretic result bears his name. Along with Cantor and Felix Bernstein (1878–1956) he independently discovered the Schröder-Bernstein Theorem which asserts that sets A and B are of equal cardinality if there exists a one-to-one map from A into B and another from B into A.

Schröder's work in the study of iteration, like his work in logic and set theory, was not fully appreciated by his contemporaries. Cayley was evidently unaware of Schröder's work and made no reference to him. Although many of those who studied iteration in the late nineteenth century were familiar with him, as is indicated by the fact that the Schröder functional equation bears his name, little was said regarding his investigation of Newton's method. For example, both Gabriel Koenigs [1884:s40–41], the central figure in the nineteenth century study of iteration, and Julia [1918:232] refer only to Cayley when discussing Newton's method.

The lack of recognition afforded Schröder's accomplishments with respect to Newton's method continues in much of the present day literature. Schröder's work regarding certain generalizations of Newton's method has, however, caught the eye of some contemporary numerical analysts (see, for example, Householder [1970]).

Schröder's research into the iteration of complex functions is predicated upon his particular conception of Newton's method. Where previously the application of Newton's method to the solution of the equation $f(z) = 0$ had been generally regarded as the discrete, real-valued numerical algorithm given above at (1.1), Schröder viewed it as the iteration of the complex analytic function

$$N(z) = z - \frac{f(z)}{f'(z)},$$

on a neighborhood of a root of the function $f(z)$.[1]

While this distinction may seem unnecessary to some contemporary readers, it is actually quite important. In order to gain insight into the workings of Newton's method, which Schröder realized was tantamount to the iteration of a particular analytic function, namely, $N(z)$, Schröder felt that it would behoove him to study the iteration of arbitrary complex functions, which he did to great profit via the application of the theory of analytic functions. The insights that Schröder gained were consequently a direct result of his novel conception of Newton's method, and it is precisely because of this that his work signals the beginning of the trend which resulted in the studies of Fatou and Julia.

1.3 Schröder's Fixed Point Theorem

Schröder's examination of iteration led him to the discovery of the following fundamental result:

Theorem 1.1 (Schröder's Fixed Point Theorem) *Let $\phi(z)$ be a function which is analytic on a neighborhood of a point x which satisfies $\phi(x) = x$ with $|\phi'(x)| < 1$. Let $\phi^n(z)$ denote the nth iterate of $\phi(z)$, that is, the n-fold composition of $\phi(z)$ with itself. Then for all z in some neighborhood D of x*

$$\lim_{n\to\infty} \phi^n(z) = x.$$

Although I have stated the result in a slightly more contemporary form than did Schröder, I have in no way distorted its content, as is indicated by Schröder's own summary of the result:

> All points z in an area around x have as a limit, under the unceasing iteration of the function $\phi(z)$, the root x of the equation $\phi(z) = x$ [1870:322].[2]

A point x which satisfies $\phi(x) = x$ for a given function $\phi(z)$ is usually called a *fixed point* of $\phi(z)$. If, in addition, the fixed point x also satisfies $|\phi'(x)| < 1$, then x is called an *attracting fixed point*. A more contemporary summary of his theorem would therefore be that all points in an arbitrarily small neighborhood of an attracting fixed point x of an analytic function $\phi(z)$ converge under iteration by $\phi(z)$ to x.

---

[1]That $N(z)$ is analytic around simple roots of an analytic function $f(z)$ is trivial; as Schröder pointed out $N(z)$ is also analytic around multiple roots of $f(z)$, since these points are what would later be called removable singularities of $N(z)$, as one can easily see by considering the expression $f(z)/f'(z)$.

[2]In the interest of clarity I have changed the Schröder's functional notation to conform to my own. For example, where he used "$F(z)$," I use "$\phi(z)$."

In order to apply his fixed point theorem to Newton's method, and consequently explain why Newton's method works near a solution x of the equation $f(z) = 0$, it is necessary to show that the Newton's method function for $f(z)$,

$$N(z) = z - \frac{f(z)}{f'(z)},$$

satisfies the hypotheses of his theorem.

That any root x of $f(z) = 0$ satisfies $N(x) = x$ is trivial if x is a simple root of $f(z)$. If x is a multiple root of $f(z)$, then the singularity which then appears in the quotient $f(z)/f'(z)$ is removable, and it can be easily shown that $N(x) = x$ in this case as well.

It is also quite easy to see that the modulus of $N'(x)$ is strictly less than one. If x is a simple root then

$$N'(z) = \frac{f(z)f''(z)}{f'(z)^2} \tag{1.2}$$

and $N'(x) = 0$. If x is a root of multiplicity p, then it can be shown via direct calculation that $N'(x) = 1 - 1/p$.

Since the Newton's method function is analytic on a neighborhood surrounding the root x, Theorem 1.1 asserts that there must exist a neighborhood D of x on which iterates of the function $N(z)$ converge to x for all points in D.

At this point in his investigations, Schröder made another observation which he evidently believed was very important: the Newton's method function is by no means the only iterative root finding function for a given equation $f(z) = 0$. Any function analytic on a neighborhood of a root x of $f(z)$, which has x as an attracting fixed point satisfies the hypotheses of Theorem 1.1 and therefore converges to x under iteration on some neighborhood of x. As will be seen in the next section, Schröder showed that such functions can be constructed with ease.

Schröder's proof of his fixed point theorem relied on infinitesimal arguments. This is somewhat surprising because Schröder was a rather innovative mathematician, but it is perhaps symptomatic of his isolation from the German mathematical mainstream, since it suggests that Schröder was unaware of Karl Weierstrass' (1815–1897) rigorous delta-epsilon approach to mathematics. Although Schröder's theorem is true, as Koenigs proved in the 1880's, Schröder did not provide a fully rigorous explanation for it.[3]

The argument Schröder did provide is in essence the following. Suppose first that Taylor expansion for $\phi(z)$ about x is of the form

$$\phi(z) = x + \phi'(0)z + \cdots,$$

where $0 < |\phi'(x)| < 1$. Let $\epsilon = z - x$ and z be sufficiently close to x so that $\phi'(x)(z - x)$ "outweighs all the succeeding terms, thus one can for infinitely small ϵ

[3] Koenigs' work is discussed in Chapter 3.

set [1870:321]"

$$\phi(z) - x = \phi'(x)\epsilon.$$

Therefore

$$\begin{aligned}\phi^2(z) - x &= \phi'(x)[\phi(z) - x] \\ &= \phi'(x)[\phi'(x)\epsilon] \\ &= \phi'(x)^2\epsilon.\end{aligned}$$

If the argument above is extended to $\phi^n(z)$, then

$$\phi^n(z) - x = \phi'(x)^n\epsilon.$$

Since $|\phi'(x)| < 1$ the right hand side goes to 0 as n approaches ∞, hence

$$\lim_{n\to\infty} \phi^n(z) - x = 0,$$

and $\phi(z)$ converges to x.

Schröder's argument suggests that iteration by $\phi(z)$ near the fixed point x acts linearly, that is, it is well-approximated by the iteration of the mapping $z \mapsto \phi'(x)(z - x)$. As will be seen, much of what happened in the study of iteration in the 1880's can be viewed as an attempt to linearize the iteration of complex functions.

No doubt motivated by the fact that the derivative of the Newton's method function $N(z)$ is 0 at simple roots of $f(z) = 0$, Schröder investigated the special case where the derivative of a function $\phi(z)$ is equal to 0 at a fixed point x. For convenience the quantity $\phi'(x)$ will be referred to as the *multiplier* of $\phi(z)$ at the fixed point x. Schröder asserted that a function whose multiplier at an attracting fixed point x is 0 in general converges to its fixed point under iteration much more quickly than does a function whose multiplier at x is non-zero; moreover, the higher the order of the first non-zero derivative at x, the quicker $\phi^n(z)$ converges to x.

Schröder gave no explicit justification of these claims, but probably believed they followed immediately from the infinitesimal approximation

$$\phi(z) - x = \frac{\phi^{(m)}}{m!}(x)\epsilon^m < A\epsilon,$$

where the higher order terms $\phi(z)$ are ignored, $\epsilon = z - x$, A is constant and m is the order of the first non-zero derivative at x.

In light of the above, he defined the order of convergence of $\phi^n(z)$ to a fixed point x as follows: if $\phi^{(i)}(x) = 0$ for $0 < i < m$ but $\phi^{(m)}(x) \neq 0$, then the convergence of $\phi(z)$ under iteration is of the *mth* order.

Before closing this section, it is worthwhile to observe that Schröder's Theorem 1.1 is strictly a *local* result because it says nothing about what happens beyond the neighborhood of an attracting fixed point. As such, it set the tone for those who directly followed him since it was the early twentieth century before anything substantial was known regarding the behavior of an arbitrary point in the complex plane under iteration of a given function. One of the triumphs of the theory established by Fatou and Julia was its sweeping description of iteration away from the vicinity of attracting fixed points.

1.4 A Generalization of Newton's Method

The following quotation from the beginning of [1870] suggests that Schröder was intrigued with the notion that there were other root finding iterative functions besides the Newton's method function.

> As will be shown there is a wide, even infinite, diversity of methods, all of which possess a common character, namely, that one can begin one's calculations with an almost arbitrary number, subject to certain laws, and through an incessant series of operations [i.e., iteration] pass to the desired result, that is, come as close as one wishes to the root. ... To develop these solution methods is the subject of the following study [1870:318–19].

Armed with his fixed point theorem, Schröder turned his attention towards this "infinite diversity." His theorem asserts that if a function $M(z)$ satisfies the following two properties,

1. $M(x) = x$, where x is a simple root of a given function $f(z)$, and
2. $|M'(x)| < 1$,

then it will converge under iteration to x on some neighborhood of x, hence it can be used to solve the equation $f(z) = 0$.

Schröder's interest in generalizing Newton's method was probably motivated by his observation, discussed at the end of the previous section, that the convergence of a function under iteration on a neighborhood of a fixed point x increases with the order of the first non-zero derivative at x. Roughly speaking, this means that a function whose first i derivatives at the fixed point are 0 converges to x more quickly the larger i is. Schröder therefore sought to improve on the rate of convergence of Newton's method for the equation $f(z) = 0$ by producing a family of functions $N_m(z)$ which satisfy the following conditions:

1. $N_m(x) = x$ where $f(x) = 0$, and
2. $N_m^{(i)}(x) = 0$ for $i < m$, and $N^{(m)}(x) \neq 0$.

In order to get the flavor of his construction, I will construct $N_2(z)$ for the equation $f(z) = 0$ with a simple root x. Set

$$N_2(z) = z - f(z)\phi(z) \tag{1.3}$$

where $\phi(z)$ is an analytic function which will be chosen in such a way to ensure that $N_2'(x) = 0$. Since $f(x)\phi(x) = 0$, it is immediate that x is a fixed point of $N_2(z)$, hence $N_2(x) = x$ is satisfied regardless of the choice of $\phi(z)$. To find $\phi(z)$ so that $N_2'(x) = 0$ is also satisfied, take derivatives of both sides of equation (1.3), set z equal to the fixed point x and solve the resulting equation for $\phi(x)$:

$$N_2'(x) = 1 - f(x)\phi'(x) - f'(x)\phi(x) = 0. \tag{1.4}$$

Since $f(x) = 0$ but $f'(x) \neq 0$, the above equation implies that

$$\phi(x) = \frac{1}{f'(x)}. \tag{1.5}$$

Thus any function $\phi(z)$ satisfying both $\phi(x) \neq 0$ and (1.5) will produce a function $N_2(z)$ with the desired properties. Of course, if $\phi(z) = 1/\,f'(z)$ then $N_2(z)$ is just Newton's method.

The computations involved in finding $N_m(z)$ increase in complexity as m grows. In a remarkable display of computational agility, Schröder found a series form for $N_m(z)$ [1870:330] as well as a formal series expression for $\lim_{m\to\infty} N_m(z)$ [1870:329].[4] Since the expression of this series requires some rather involved notation of Schröder's own invention, it will be omitted here. As an indication of the complexity of these expressions, Schröder's generalized Newton's method $N_6(z)$ is [1870:330]:

$$\begin{aligned} z &- \frac{f}{1!}\frac{1}{f'} - \frac{(f)^2}{2!}\frac{f''}{(f')^3} - \frac{(f)^3}{3!}\frac{3(f'')^2 - f'f'''}{(f')^5} - \frac{(f)^4}{4!}\frac{15(f'')^3 - 10f'f''f''' + (f')^2 f^{(iv)}}{(f')^7} \\ &- \frac{(f)^5}{5!}\frac{105(f'')^4 - 105f'(f'')^2 f''' + 10(f')^2(f''')^3 + 15(f')^2 f'' f^{(iv)} - (f')^3 f^{(v)}}{(f')^9}. \end{aligned} \tag{1.6}$$

Schröder followed his first generalization of Newton's method with two more families of iterative algorithms, which he denoted $\mathcal{A}$ and $\mathcal{B}$.[5] He generated $\mathcal{A}$ from the Newton's method function $N(z)$, and $\mathcal{B}$ from yet another generalization of Newton's method, the function $M(z)$, where $M(z)$ is the Newton's method function for $f(z)/f'(z) = 0$, that is

$$M(z) = z - \frac{f(z)}{f'(z)}\left(\frac{f(z)}{f'(z)}\right)' = z - \frac{f(z)f'(z)}{f'(z)^2 - f(z)f''(z)}. \tag{1.7}$$

[4]Schröder indicated that, as $m \to \infty$, the function $N_m(z)$ approaches the constant function $G(z) \equiv x$ at x. The real question is whether the sequence $\{N_m(z)\}$ actually converges to the constant function $G(z) \equiv x$ on a neighborhood of the root. Although Schröder addressed this issue in the special case where $f(z)$ is a quadratic polynomial, he had little to say about this matter in the general case.

[5]These families are of interest in certain areas of numerical analysis, see for example Householder [1970].

The term $f(z)/f'(z)$ has the effect of converting the multiple roots of $f(z)$ to simple ones, hence $M(z)$ is a variant of Newton's method which is particularly useful when $f(z)$ has multiple roots. In this instance $M(x) = x$ and $M'(x) = 0$ regardless of the multiplicity of the root, thus the convergence under iteration of $M(z)$ to x is, like that of $N(z)$ in the case of a single root, of the second order. Moreover, the function $M(z)$ plays a pivotal role in Schröder's examination of the convergence of Newton's method for the complex quadratic, which will be discussed below.

A second element of Schröder's treatment of Newton's method for the quadratic is the idea of a conjugacy, which he explored in the second of his two papers on iteration, the paper [1871]. Where his first paper [1870] concerned the study of iterative root solving functions such as the generalizations of Newton's method discussed above, Schröder in his second paper delved more deeply into the study of iteration of arbitrary complex analytic functions.

1.5 Schröder's Paper [1871]

Having established the motivation for the study of iteration in his first paper—namely, the study of equation solving functions such as Newton's method—Schröder turned his energies in his second paper, [1871], towards the study of iteration in general. He evidently felt, with some justification, that he was a mathematical pioneer, for he began his second paper with the following remark:

> I consider herein the study of a field in which I have encountered very few collaborators [1871:296].

One of his principal concerns in this paper was the practical problem of how to represent the *nth* iterate of a given function $\phi(z)$. He suggested that one method of doing this would be to derive a direct formula for its *nth* iterate. In general he realized that obtaining such a formula would be difficult, if not impossible.

However, he did produce such a formula (see (1.9) below) for the linear fractional transformation by modifying a similar formula for the *nth* iterate of $L(z)$ found in the famous textbook *Cours D'Algébre Supèrieure* written by the French mathematician Joseph Alfred Serret (1819-85).[6]

In order to present Schröder's version of this formula, let

$$L(z) = \frac{az+b}{cz+d} \tag{1.8}$$

[6] Serret's formula for a linear fractional transformation was evidently not motivated by an interest in the general properties of iteration. That it followed a discussion of the group theoretic properties of substitutions on n letters suggests that he was interested in exploring similarities between groups of linear fractional transformations and substitutions of n letters [Serret 1866/1885:356–57]. Not only did Serret's formula give him the precise form of the *nth* iterate of a linear fractional transformation, it allowed him to determine whether the *nth* iterate of a given linear fractional transformation $L(z)$ equals the identity function.

have distinct fixed points ξ_1 and ξ_2. Schröder showed that [1871:299]

$$L^n(z) = \frac{(\xi_1 + \frac{d}{c})^n(\xi_1 z + \frac{b}{c}) - (\xi_2 + \frac{d}{c})^n(\xi_2 z + \frac{b}{c})}{(\xi_1 + \frac{d}{c})^n(z - \xi_2) - (\xi_2 + \frac{d}{c})^n(z - \xi_1)}. \tag{1.9}$$

Reflective of his interest in finding the limit of an iterative process, Schröder deduced that if

$$|\xi_1 + \frac{d}{c}| > |\xi_2 + \frac{d}{c}|, \tag{1.10}$$

which is equivalent to assuming that ξ_1 is the attractive fixed point, then

$$\lim_{n\to\infty} L^n(z) = \frac{\xi_1 z + \frac{b}{c}}{z - \xi_2}. \tag{1.11}$$

Equation (1.11) leads to an interesting result concerning the convergence of the iterates of $L(z)$ which, surprisingly, Schröder neglected to investigate. According to his fixed point theorem, the fact that ξ_1 is an attracting fixed point implies that there is a neighborhood of ξ_1 on which $L(z)$ converges under iteration to ξ_1. Setting

$$\lim_{n\to\infty} \frac{\xi_1 z + \frac{b}{c}}{z - \xi_2} = \xi_1$$

and solving for z to find the values for which z converges to ξ_1 under iteration, yields the identity

$$-\xi_1\xi_2 = \frac{b}{c}. \tag{1.12}$$

From this it follows that under the conditions stated above at equation (1.10) $L(z)$ converges to the fixed point ξ_1 on the entire plane save the fixed point ξ_2.

Given Schröder's interest in the convergence properties of a function under iteration, it seems odd that he did not take note of this fact, for it would have provided an interesting example of a family of functions which converge to a single attracting fixed point on virtually the entire plane. It is possible that his intention in presenting formula (1.9) was only to provide an example of a function whose *nth* iterate could be nicely expressed.

1.6 Schröder and Functional Equations

Schröder's belief that in general he would not likely find a formula for the *nth* iterate of a given function $\phi(z)$ led him to consider the notion of conjugation. A function $\phi(z)$ is said to be analytically conjugate to a function $\psi(z)$ on a disc D if there exists an analytic function $F(z)$ such that the following diagram commutes:

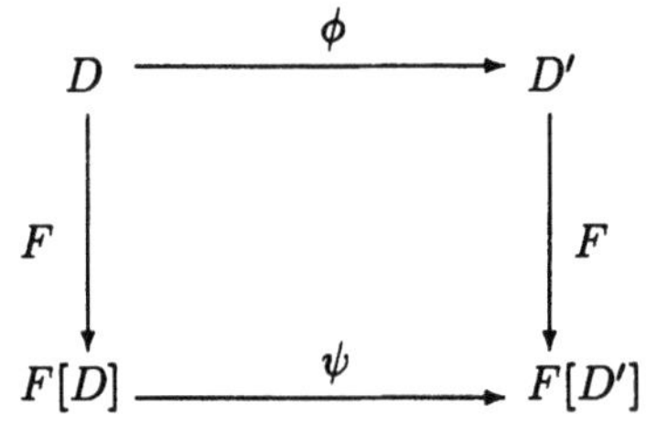

Many important iterative properties, such as convergence to a fixed point under iteration, are preserved by conjugation. For example, if x is a fixed point of $\phi(z)$, then $F(x)$ is a fixed point of $\psi(z)$. If $F(z)$ is also invertible on D then the diagram implies that

$$\phi^n(z) = F^{-1} \circ \psi^n \circ F(z),$$

which reduces the iteration of $\phi(z)$ to that of $\psi(z)$. If $\psi(z)$ is in addition easier to iterate than $\phi(z)$, then this reduction greatly simplifies the study of $\phi(z)$ under iteration. For this reason, the notion of conjugation is of great use in the study of complex dynamics.

Schröder's discovery of conjugation was prompted by his desire to find a function $\psi(z)$ which is not only easier to iterate than a given function $\phi(z)$ but, more importantly, to which the iteration of $\phi(z)$ easily reduces.

That Schröder had the concept of conjugation in mind is made clear by his posing of the following problem [1871:310]: given a function $\phi(z)$, find an invertible function $G(z)$ and a function $\psi(z)$ which satisfies

$$\phi(z) = G \circ \psi \circ G^{-1}(z). \tag{1.13}$$

In a detail lacking elsewhere in his work, which suggests he was dealing with ideas he judged unfamiliar to his audience, he carefully demonstrated the trivial fact that if (1.13) is satisfied by a function $\psi(z)$ then

$$\phi^n(z) = G \circ \psi^n \circ G^{-1}(z). \tag{1.14}$$

Schröder was not the first mathematician to consider the notion of conjugation. The British mathematician Charles Babbage (1792–1871) did so much earlier in his paper [1820]. However, Babbage's study of functional equations was not "dynamic" in the sense that he did not consider notions such as that of an attracting fixed point, nor was his interest in functional equations motivated by an interest in iteration. Moreover, there is no evidence that Schröder drew upon Babbage's study of functional equations, and it is quite likely that Schröder came to the notion of conjugation independently. Nonetheless, there was some interest in Babbage's work among those who contributed to the subsequent development of complex dynamics (see the paper [Leau 1898], which concerns the so-called Babbage functional equation, $F^n(z) = F(z)$).

As examples of functions $\psi(z)$, he suggested $\psi(z) = zh$ or $\psi(z) = z + h$, where h is a complex constant. If equation (1.14) is satisfied by $\psi(z) = z + h$ then

$$\phi^n(z) = G(G^{-1}(z) + nh),$$

and iteration reduces to repeated addition of h. If, on the other hand, $\psi(z) = hz$, then iteration reduces to repeated multiplication by h since [1871:303]

$$\phi^n(z) = G(h^n G^{-1}(z)).$$

Schröder therefore reduced the iteration of an arbitrary given function $\phi(z)$ to that of either $z \mapsto hz$ or $z \mapsto z + h$ via the solution of either of the following functional equations:

$$\phi(G(z)) = G(z + h) \tag{1.15}$$

or

$$\phi(G(z)) = G(hz). \tag{1.16}$$

The functional equations (1.15) and (1.16) are nowadays usually stated in the equivalent forms

$$F(\phi(z)) = F(z) + h \tag{1.17}$$

and

$$F(\phi(z)) = hF(z), \tag{1.18}$$

where $F(z) = G^{-1}(z)$, and are generally referred to, respectively, as the Abel and Schröder functional equations.[7]

The existence of a function $F(z)$ solving the Abel equation implies that the following diagram commutes:[8]

---

[7]The Abel functional equation is found in Abel [1824?], a somewhat obscure, posthumous fragment of uncertain date, included in the collected works of Niels Abel published in 1881 (see [1881,II:37]). Abel's treatment of the so-called Abel equation is discussed in detail in the next chapter. It is doubtful that his interest in this equation stemmed from an interest in the iteration of arbitrary functions. It is also doubtful that Schröder was familiar with Abel [1824?]. Not only does he not cite it, but there were evidently very few copies of this paper in circulation before 1881. In any event, Schröder's treatment of this equation had nothing in common with that of Abel.

[8]The use of a commutative diagram is not meant to suggest that Schröder, or anyone else whose work is discussed in this study, used commutative diagrams. Although the implicit use of conjugacies via the functional equations given by Schröder became pervasive in the study of the iteration of complex analytic functions, such conjugacies were usually expressed as functional equations and not, until recently, as commutative diagrams.

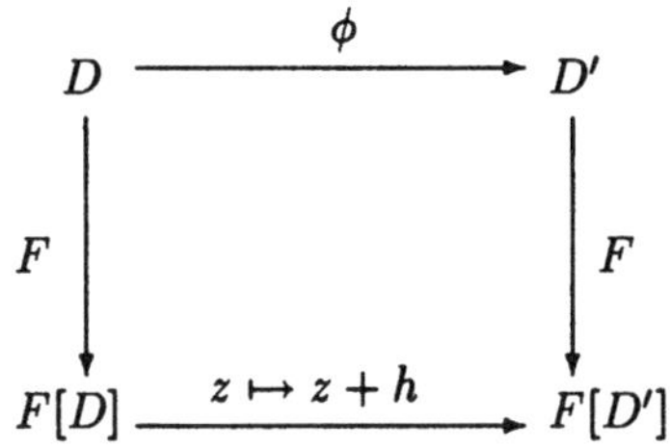

A solution of the Schröder equation would likewise provide a conjugacy between $\phi(z)$ and $z \mapsto hz$.

Both of these functional equations are fundamental in the contemporary study of complex dynamics, and a particular version of the Schröder equation at (1.18),

$$F(\phi(z)) = \phi'(x)F(z)$$

where, $\phi(x) = x$ and $0 < |\phi'(x)| < 1$, has become one of the most widely used functional equations in complex dynamics. The existence of a solution $F(z)$ on a neighborhood of the fixed point x to this version of Schröder's equation implies that $\phi(z)$ is conjugate to multiplication by its derivative in the neighborhood of a fixed point, which suggests in turn that near an attracting fixed point of $\phi(z)$, iteration is equivalent to multiplication by the derivative at the fixed point. Although Schröder never stated this particular version of the Schröder equation explicitly, it was suggested in his proof of the fixed point theorem, Theorem 1.1 above, where he reduces the iteration of $\phi(x)$ near x to repeated multiplication of $\phi'(x)$.

Schröder had great difficulty developing a general approach to the solution of functional equations (1.17) and (1.18). The instances where he solved either the Abel or the Schröder equations for a particular function $\phi(z)$ amount to little more than special case solutions and do not involve methods which generalize to the solution of either equation for arbitrary functions. Schröder readily conceded these shortcomings:

> In general, given functions $\phi(z)$ and $\psi(z)$, the search for the function $F(z)$ [that is, a solution to the Abel or Schröder equation] is a difficult task, and one should instead take the opposite approach [1871:302].

Schröder's "opposite approach" was to fix an arbitrary function $\psi(z)$ and then consider functions of the form

$$F^{-1} \circ \psi \circ F(z)$$

as $F(z)$ varies. With a bit of luck, an interesting function $\phi(z)$ may emerge which, for a given $F(z)$, satisfies

$$\phi(z) = F^{-1} \circ \psi \circ F(z),$$

which then implies that

$$F(\phi(z)) = \psi(F(z)).$$

For example, if $\psi(z) = 2z$ and $F(z) = \arctan(iz)$, where $\arctan(z)$ is a fixed inverse of the tangent function, Schröder showed that

$$\frac{2z}{1+z^2} = -i\tan(2\arctan(iz)). \tag{1.19}$$

Schröder generated (1.19) from the well-known trigonometric identity

$$\frac{2z}{1-z^2} = \tan(2\arctan(z)), \tag{1.20}$$

by letting $z \mapsto iz$ in both sides of equation (1.20).

It will be seen later that the French mathematician Samuel Lattès (1873–1918), whose work will be discussed in Chapter 10, approached the solution of a particular functional equation along lines reminiscent of Schröder's "opposite approach."

1.7 Schröder and Newton's Method for the Quadrati

Schröder proved the following theorem regarding the convergence of Newton's method in the paper [1871].

Theorem 1.2 (Schröder) *Let $q(z)$ be a complex quadratic with distinct roots. Let*

$$N(z) = z - \frac{q(z)}{q'(z)}$$

be the Newton's method function for $q(z)$, and let L be the perpendicular bisector of the line segment connecting the roots of $q(z)$. Let H_1 and H_2 be the open half-planes into which the line L divides $\bar{\mathbb{C}}$. If α_1 is the root of $q(z)$ in H_1 and α_2 the root in H_2, then for all z in H_i, Newton's method converges under iteration to α_i, that is,

$$\lim_{n\to\infty} N^n(z) = \alpha_i.$$

For all z on L, however, $N(z)$ does not converge under iteration to either α_1 or α_2.

He proved the above theorem indirectly, since he viewed it as a consequence of the following theorem with which he concluded his second paper in iteration, the paper [1871]:

Theorem 1.3 (Schröder) *Let $q(z) = z^2 - 1$. For $q(z)$ the generalized Newton's method function*

$$z - \frac{q(z)q'(z)}{(q'(z))^2 - q(z)q''(z)} \tag{1.21}$$

is the function

$$M(z) = \frac{2z}{z^2+1}.$$

On the left open half-plane $M(z)$ converges under iteration to -1. On the right open half-plane $M(z)$ converges to 1, while on the imaginary axis, $M(z)$ converges to neither root.[9]

The reason that Schröder reduced Theorem 1.2 to Theorem 1.3 involves the fortuitous circumstance discussed at the end of the previous section, namely, that

$$M(z) = \frac{2z}{1+z^2} = -i\tan(2\arctan(iz)).$$

This in turn implies that

$$M^n(z) = -i\tan(2^n \arctan(iz)). \tag{1.22}$$

After a long calculation which drew upon algebraic facts regarding the application of Newton's method which he had established in his previous paper [1870], Schröder showed that

$$\lim_{n\to\infty} -i\tan(2^n \arctan(z)) = 1$$

for z on the right half-plane and -1 for z on the left half-plane.

Schröder was then able to reduce the iteration of $N(z)$ to that of $M(z)$ because both of these functions share the property that

$$\phi(z) = \phi\left(\frac{1}{z}\right). \tag{1.23}$$

Such a function has the unusual property that it is conjugate to its reciprocal, which is seen by observing that the following diagram commutes:

---

[9] As was mentioned at the end of Section 1.4 at equation (1.7), Schröder defined $M(z)$ for arbitrary functions in the same manner as he did for the function $q(z) = z^2 - 1$ in the above theorem. Along with the Newton's method function $N(z)$, he used it to generate two families of iterative, root-finding functions, $\mathcal{A}$ and $\mathcal{B}$.

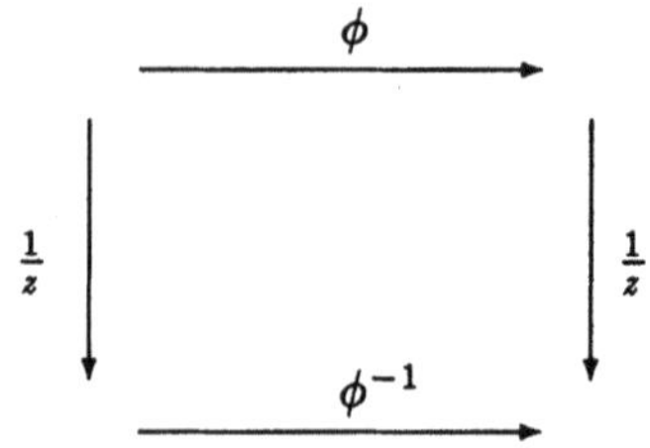

Thus, since

$$N(z) = \frac{1}{M(z)}$$

it follows that $M(z)$ and $N(z)$ are conjugate to one another, hence

$$\lim_{n\to\infty} N^n(z) = \lim_{n\to\infty} \frac{1}{M^n(z)}.$$

As noted above,

$$\lim_{n\to\infty} M^n(z) = \pm 1,$$

the sign depending on the half-plane from which z is taken, and consequently

$$\lim_{n\to\infty} N^n(z) = \pm 1,$$

where, again, the sign depends on z.

Despite the good fortune that, for $q(z)$, the generalized Newton's method function $M(z)$ happens to be

$$-i\tan(2\arctan(iz)),$$

which in turn is the inverse of $N(z)$, Schröder's proof is impressive, and his reduction of the iteration of $M(z)$ to repeated multiplication by the constant 2 indicates that his faith in iteration as a mathematical tool was not misplaced.

Schröder followed his evaluation of this limit with a brief investigation of the iteration of $M(z) = 2z/(z^2+1)$ on the set of points which converge to neither root, which happens to be the imaginary axis. Using the form of $M(z)$ given in equation (1.19) he showed that if $M^n(z) = z$ with $n > 1$, then z must be of the form

$$z = i\tan\left(\frac{k\pi}{2^n - 1}\right).$$

He observed as well that there exist points z on the imaginary axis such that the set $S = \{M^n(z)\}$ takes on infinitely many values yet has no discernible order.

Consistent with the virtual non-existence of a theory of sets in 1870, Schröder made no topological observations regarding either the set S or the set of points which satisfy $M^n(z) = z$.

Before discussing Cayley's treatment of Theorem 1.2, which is quite different from Schröder's, a few words concerning Schröder's proof are in order. First of all, Schröder's justification that the special case $q(z) = z^2 - 1$ is representative of the general case occurs at the end of his first paper. Rather than supply it, I will note only that it is not too hard to show that the Newton's method functions for any two quadratics with distinct roots are linearly conjugate. Since convergence to a fixed point is a property preserved by conjugation, the argument in the special case of $q(z) = z^2 - 1$ implies the general case.

It is also important to bear in mind that Schröder wrote in the leisurely and discursive style prevalent in pre-twentieth century mathematics. This style certainly has its charms, but, at least in Schröder's case, it can also present difficulties. For example, these proofs were by no means executed in an orderly fashion. The various stages are interspersed throughout both [1870] and [1871]. During the final phase of his proof of Theorem 1.3 he at times seems to pull things out of nowhere, but upon reflection it is realized that Schröder is drawing on something he had considered in an earlier section of his work.

These sorts of problems are more of an annoyance than a particularly major flaw, but it can make his line of thought difficult to follow. Cayley's treatment of Theorem 1.2 is much more concise and, as was noted at the outset, is often cited as the first proof of this theorem, although it appeared almost ten years after Schröder's. This could be attributed to many things, including the fact that mathematical papers often did not circulate widely in Schröder's day, but it is also possible that the sort of stylistic problems just discussed account for it. That many of the nineteenth century mathematicians who studied iteration were aware of some aspects of his work, but apparently not Theorem 1.3, suggests such a possibility.

It is a bit ironic that Cayley's proof of Theorem 1.2 is nowadays generally better known than Schröder's, since the latter's view of Newton's method as the iteration of an analytic function near an attracting fixed point is much closer to the contemporary spirit than is the view held by Cayley, who adhered to the traditional conception of Newton's method as a discrete process. Consequently, Cayley did not connect Newton's method to the theory of functions, which perhaps explains why Cayley's interest in Newton's method did not engender a concomitant investigation of the process of iteration, as it did with Schröder.

1.8 Arthur Cayley and Newton's Method

Cayley presented two proofs of Theorem 1.2, a geometrically flavored proof in his paper [1879a] and an algebraic refinement of his earlier proof which appeared in *Comptes rendus* of the French Academy of Sciences in 1890. He also wrote two short notes [1879b] and [1880], where he discussed his investigation of Newton's method. Aside from differences in approach there are no substantial differences in Cayley's two proofs of Theorem 1.2. Because of its intriguing geometric flavor I will, in what follows, focus on his proof from his paper [1879a].

Like Schröder, he felt free when convenient to reduce the case of the general quadratic to that of $q(z) = z^2 - 1$. Moreover, Cayley's treatment of the quadratic case was ad hoc— as was Schröder's in some respects—since his methods generalized neither to the study of iteration nor to the study of Newton's methods for higher degree polynomials.

In the introduction to his 1879 proof of Theorem 1.2, Cayley derived Newton's method for the special case $p(z) = z^2 - a^2$, where a is complex. His derivation is reminiscent of the sort of non-rigorous use of infinitesimals favored by Schröder. Cayley began by choosing an initial approximation z_0 of the root a and set $a = z_0 + h$, where h is a small quantity. He obtained the approximation

$$h = -\frac{z_0^2 - a^2}{2z_0},$$

by expanding the expression

$$a^2 = (z_0 + h)^2,$$

ignoring the higher order term h^2 and then solving for h. Adding z_0 to both sides of the above expression for h and writing $z_1 = z_0 + h$, he produced the first approximation z_1:

$$\begin{aligned} z_1 &= z_0 + h \\ &= z_0 - \frac{z_0^2 - a^2}{2z_0} \\ &= z_0 - \frac{p(z_0)}{p'(z_0)}. \end{aligned} \tag{1.24}$$

He concluded by remarking that

> ...the question is, under what conditions do we thus approximate to one determinate root (selected out of the roots at pleasure), say a, of the given equation [1879a:114].

Cayley's answer came in the form of a geometric construction of Newton's method in the quadratic case, which he also justified algebraically in the second half of his paper. For the geometric phase of his proof, Cayley used the quadratic $p(z) = z^2 - a^2$; in the algebraic phase he switched to $q(z) = z^2 - 1$. I will present his geometric construction for Newton's method in the case where $q(z) = z^2 - 1$.

Cayley's examination of Newton's method for the quadratic began with the construction of a family of circles, C_k, where k is a positive real constant. He defined the circle C_k as the set of points z satisfying

$$k = \frac{|z-1|}{|z+1|}, \tag{1.25}$$

which is the circle

$$\left(x - \frac{1+k^2}{1-k}\right)^2 + y^2 = \frac{4k^2}{(1-k^2)^2} \tag{1.26}$$

where $z = x + iy$. The radius and center of C_k are

$$\left|\frac{2k}{1-k^2}\right| \text{ and } \left(\frac{1+k^2}{1-k^2}, 0\right). \tag{1.27}$$

The heart of Cayley's geometric representation is his demonstration that Newton's method carries the circle C_k to the circle C_{k^2}. Once this was done, he picked an arbitrary z from the right half-plane, which forces $k < 1$ since the segment connecting z to 1 is shorter than the segment connecting z to -1, and constructed the sequence of circles $\{C_{k^{2^n}}\}$, which represents successive applications of Newton's method. In an informal induction, he concluded that since Newton's method carries C_k to C_{k^2}, it also carries the circle $C_{k^{2^n}}$ to $C_{k^{2^{n+1}}}$.

Equation (1.26) implies that the radius and center of this sequence of circles respectively approach 0 and $(1,0)$ as n goes to infinity, consequently Newton's method limits on the root $z = 1$. Similar arguments apply to points in the left plane to the root $z = -1$, and, as Cayley noted in a later paper, points on the imaginary axis do not converge to either root [1890:897].

Cayley intended his description of Newton's method for the quadratic to be the first step of an investigation into the convergence properties of Newton's method for polynomials of arbitrary degree. Cayley's plan, however, stalled with the degree three case. Unfortunately, nowhere in his published writings did he specify the sort of problems he faced except to remark that

> ... the problem is to divide the plane into regions, such that, starting with a point P_1 anywhere in one region we arrive at the root A; anywhere in another region we arrive ultimately at the root B; and so on for the several roots of the equation. The division is made without difficulty in the case of the quadratic; but in the succeeding case, that of a cubic equation, it is anything but obvious what the division is: and the author has not succeeded in finding it [1880:143].

It is not too difficult to speculate on the nature of the difficulties Cayley encountered. Let $c(z)$ denote a cubic with three distinct roots, and let $N_c(z)$ denote the corresponding Newton's method function. Fatou and Julia showed some forty years subsequent to Cayley's investigation that, unlike the quadratic case, where the imaginary axis divides the plane into two convergence regions, the set of points which divides the convergence regions of $N_c(z)$ is an extremely complicated fractal curve which partitions the extended complex plane $\bar{\mathbb{C}}$ into infinitely many components. Moreover, in any neighborhood of a point from this curve, there are points which will converge to each root of $c(z)$.

Cayley no doubt experimented with the cubic and probably expected that in the general degree n case there would be finitely many convergence regions which depended on n in an orderly fashion. However, the difficulties outlined in the previous paragraph would have made it virtually impossible for Cayley to decipher the behavior of Newton's method for higher degree polynomials via direct calculation.

Chapter 2

The Next Wave: Korkine and Farkas

2.1 Introduction

Schröder developed several notions which are central to the study of complex dynamics. Despite his failure to rigorously establish his fixed point theorem, it is a fundamental result, and his belief that iteration of an arbitrary function $\phi(z)$ could be reduced to the solution of the so-called Abel and Schröder functional equations was prophetic. Not only was the next phase in the development of complex dynamics ushered in by an interest in the solution of the Schröder and Abel equations, but the study and solution of functional equations is fundamental in many contemporary studies of iteration.

Several papers addressing the same issues Schröder raised appeared in various Parisian mathematical journals in the early 1880's. The first of these papers was written by the Russian mathematician Alexandr Korkine and appeared in 1882 in the *Bulletin des Sciences Mathématiques et Astronomiques*, also known as *Darboux's Journal*. Korkine's article was followed by a paper from the Hungarian Jules Farkas (1847–1915), [1884], which was published in *Liouville's Journal*, the *Journal de Mathématiques Pure et Appliquées*, and by three papers written by the French mathematician Gabriel Koenigs, [1883], [1884] and [1885]. The first of Koenigs' articles was also published in *Darboux's Journal*, the other two in the *Annales de L'École Normale*. Korkine's and Farkas' work will be discussed in this chapter, Koenigs' in the next.

Besides a shared interest in functional equations, these papers also evidenced an interest in a problem that Schröder raised concerning the definition of non-integer iterates of a function, that is, the definition of the *wth* iterate of an analytic function

$\phi(z)$, where w is a non-integer real, or even complex, number. To distinguish non-integer iterates from integer iterates, the letter w will be used when the index of iteration is not necessarily an integer.[1] The generalization of iteration to non-integer values is often called continuous iteration or the problem of analytic iteration.

Although the subject of analytic iteration is not in the mainstream of contemporary complex dynamics, there is some interest in the subject in contemporary mathematics (see, for example, [Erdős-Jabotinsky 1960] and [Kuczma 1990]). However, the problem of analytic iteration interested most of the nineteenth century mathematicians who investigated the iteration of complex functions. Although my interest in Korkine and Farkas lies in their treatment, respectively, of the Abel and Schröder functional equations, the problem of analytic iteration was a principal concern of both men.

Before reviewing the responses of Korkine and Farkas to Schröder's study of functional equations it will be useful to first say a few words about analytic iteration, and then to briefly outline the respective approaches of Schröder, Korkine and Farkas to this problem.

2.2 Analytic Iteration

Given an arbitrary analytic function $\phi(z)$,the problem of analytic iteration is to find a function $\Phi(w, z)$ from $A \times \bar{\mathbb{C}}$ to $\bar{\mathbb{C}}$, where A is either real or complex, which is analytic in the complex variable z, continuous in A and satisfies the following two conditions:

$$\Phi(w+u, z) = \Phi(w, \Phi(u, z)) \tag{2.1}$$
$$\Phi(1, z) = \phi(z). \tag{2.2}$$

For positive integer w, such a function can always be found, namely, the function

$$\Phi(w, z) = \phi^w(z).$$

In fact, condition (2.2), together with condition (2.1), implies that if $\Phi(w, z)$ exists, then $\phi^w(z) = \Phi(w, z)$ for positive integer w. If a function $\Phi(w, z)$ can be found which satisfies both the conditions at (2.1) and (2.2) for non-integer or even complex w, then the *wth* iterate of $\phi(z)$ can be defined as

$$\phi^w(z) = \Phi(w, z).$$

A function $\Phi(w, z)$ satisfying the conditions at (2.1) and (2.2) for a given analytic function $\phi(z)$ will be referred to as an *analytic iteration function*; the *problem of analytic iteration* will refer to the attempt to find such a function $\Phi(w, z)$.

[1] Although the situation did not arise in the previous chapter, it is quite common to speak of the *nth* iterate when n is a negative integer. In such a case $\phi^n(z)$ denotes, depending on the context, either the *nth* iterate of a particular inverse of $\phi(z)$, or the set of the *nth* preimages of $\phi(z)$.

Schröder proposed finding such a function, where w and z are both complex, in the beginning of his second paper on iteration, the paper [1871]:

> [For a given analytic function $\phi(z)$] find a function $\Phi(w, z)$ of two complex arguments w and z, which is continuous in both the w plane [and] in the z plane, [and] which furthermore satisfies the functional equation
>
> $$\Phi(w, z) = \Phi(w - 1, \phi(z)) \tag{2.3}$$
>
> including the initial condition $\Phi(1, z) = \phi(z)$ [1871:298].[2]

Schröder's version of the problem of analytic iteration is thus similar to the one given above, except that in the former, (2.3) is replaced by the more general condition (2.1).

As will be shown shortly, Schröder in effect showed that if $\Phi(w, z)$ can be defined for a positive real number w, then there is a canonical continuous curve which contains the integer iterates of $\phi(z)$. Indeed, this seems to be one of the principal reasons that Schröder introduced the problem of complex iteration, although he also seemed interested in the formal aspects of the problem.

Immediately after completing his proof of Theorem 1.3, discussed in the previous chapter, which asserts that the function $M(z) = 2z/(z^2 + 1)$ converges under iteration to the roots of $q(z) = z^2 - 1$ except on the imaginary axis, Schröder defined a curve on which, for a given z, the integer iterates of $M(z)$ reside.

In order to define this curve it is helpful to recall from the discussion preceding Theorem 1.2 that

$$M(z) = -i\tan(2\arctan(iz)),$$

which in turn implies that the *nth* integer iterate of $M(z)$ is

$$M^n(z) = -i\tan(2^n\arctan(iz)) \tag{2.4}$$

(see equation (1.22)). Schröder fixed z in (2.4) and let n take on positive real values, which generates what he termed a "continuous curve" on which lie "the consecutive approximations" of $M(z)$.

Although he offered no explicit expression for this curve, it was defined implicitly as the function

$$\Phi(w, z) = -i\tan(2^w\arctan(iz)) \tag{2.5}$$

with z fixed and w restricted to positive real values, in which case positive iterates of $M(z)$ would be of the form

$$M^w(z_0) = -i\tan(2^w\arctan(iz_0)).$$

[2] Although Schröder used the word "stetig", which translates as continuous, to describe the iteration function he sought, it is apparent from the context that Schröder sought an iteration function which was actually analytic, that is, has a power series expansion. This suggests that Schröder, like almost all mathematicians at the time, did not distinguish sharply between continuous and differentiable functions.

Schröder then proceeded to let both z and w vary, with w still real, and described the set of curves thus produced as a "peculiar fabric or net of curves [1871:321]." Although he did not return to it in [1871], he suggested that this set of curves was worthy of closer investigation.

Schröder gave other examples of non-integer iteration. It is surprising, however, that he did not explicitly state a general method for finding the analytic iteration function based on the solution of functional equations, since his construction of the curve defined by equation (2.5) suggests such an approach. Suppose that for a given analytic function $\phi(z)$ an invertible complex analytic function $F(z)$ is found which satisfies the Schröder functional equation

$$F(\phi(z)) = hF(z),$$

where h is a complex constant. Then the analytic iteration function $\Phi(w, z)$ can be defined as follows:

$$\Phi(w, z) = F^{-1}(h^w F(z)). \tag{2.6}$$

An invertible complex analytic solution $f(z)$ of Abel's functional equation

$$f(\phi(z)) = f(z) + h,$$

if one exists, likewise yields a function $\Phi(w, z)$, which can be defined formally as

$$\Phi(w, z) = f^{-1}(f(z) + wh). \tag{2.7}$$

Korkine and Farkas in fact took this very approach. Korkine reduced the problem of analytic iteration to the solution to the Abel equation in the manner suggested immediately above, and Farkas in his paper [1884] likewise used solutions of the Schröder equation to define an analytic iteration function. Both approaches were, however, inherently flawed.

Korkine offered two constructions of the analytic iteration function $\Phi(w, z)$.[3] The difficulty with each is rooted in a tendency to assume the existence of certain functions without first proving that they do in fact exist. I will discuss his work in more detail in Section 2.5.

In his paper [1884] Farkas rigorously proved the existence of a convergent power series solution to the Schröder equation

$$F(\phi(z)) = \phi'(x)F(z),$$

on a neighborhood of x provided the complex analytic function $\phi(z)$ satisfies the fixed point conditions $\phi(x) = x$ and $0 < |\phi'(x)| < 1$, as well as a peculiar set of

[3]Korkine did not specify the precise nature of the variables w and z. He remarked, in effect, that there were no restrictions of the variable w, and from context he seems to treat both $f(z)$ and $\phi(z)$ as complex functions.

conditions dictated by his proof (see Theorem 2.1 below). Farkas then used his solution $F(z)$ to the Schröder equation to define the analytic iteration function

$$\Phi(w,z) = F^{-1}(h^w F(z))$$

where w is real and z is complex on the neighborhood of the fixed point x.

Farkas' definition of the function $\Phi(w,z)$ is equivalent in spirit to the analytic iteration function

$$\Phi(w,z) = -i\tan(2^w \arctan(iz))$$

implicitly defined by Schröder and given above at equation (2.5). That Schröder may have influenced Farkas' approach is indicated by the following quotation:

> From $[F(\phi(z)) = hF(z)]$ one can easily deduce that
>
> $$F(\phi^w(z)) = h^w F(z),$$
>
> a formula Schröder suggested as well, but without realizing its generality, in his memoir on iterative functions [1884:107].

Farkas' comment regarding Schröder's alleged unawareness (which should, by the way, be taken with several grains of salt) probably refers to the fact, noted above, that Schröder did not explicitly state an approach to analytic iteration based on the solution of the Schröder equation.

In any event, Farkas' approach does not generalize completely to the case where w is complex. As Farkas knew—and as Schröder argued almost fifteen years earlier in his paper [1870]—if $\phi(x) = x$ and $0 < |\phi'(x)| < 1$, then there exists a disc D surrounding x such that for integer n,

$$\lim_{n\to\infty} \phi^n(z) = x, \tag{2.8}$$

for all z in D. Thus, if the concept of iteration is to be extended to allow for complex iterates, it is reasonable to expect that the following limit should converge to x for all z in D:

$$\lim_{w\to\infty} \phi^w(z) = \lim_{w\to\infty} \Phi(w,z) = \lim_{w\to\infty} F^{-1}(h^w F(z)).$$

However, this limit is not well-defined for complex w since $w = \infty$ is an essential singularity of the function $g(w) = h^w$.

Many of the mathematicians who followed Schröder, Korkine and Farkas treated the analytic iteration problem via the solution of the Abel or Schröder equations, and consequently either encountered the difficulties outlined above, or restricted the variable w in such a way as to avoid them.

2.3 Korkine and the Influence of Abel

Alexandr Korkine (1837–1908) figured prominently in the so-called "Petersburg School" which was a group of Russian mathematicians centered around Pafnuty Chebyshev (1821–1894) who were connected with the University of Saint Petersburg. Korkine was a pupil and mathematical disciple of Chebyshev's and received his doctorate from the University in 1867. Korkine earned renown as both an educator and a researcher. With his frequent collaborator Egor Zolotarev (1847–1871), he often ran seminars for advanced pupils. Among Korkine's students numbered many famous Russian mathematicians including Andrei Markov (1856–1922). Korkine's research interests included partial differential equations and quadratic forms. With Zolotarev, he found the upper limit for the minima of positive quadratic forms of four and five variables for a given discriminant.[4]

The impetus for Korkine's interest in both the Abel functional equation and the problem of analytic iteration came not only from the work of Schröder but also from Niels Abel (1802–1826) himself. In fact, it seems quite likely that Abel's work was a greater motivation for Korkine than was Schröder's work. That this may be the case is suggested by the fact that while Korkine featured Abel prominently in his paper [1882], he cited Schröder only fleetingly, and despite Korkine's explicit interest in the problem of analytic iteration, nowhere did Korkine mention that Schröder had first posed the problem.

One possible reason for Korkine's interest in Abel is that immediately prior to the publication of Korkine's paper [1882], Abel's work had for various reasons attracted the attention of the French mathematical community. In 1881 the *Oeuvres Complétes d'Abel*, edited by Sophus Lie (1842–1899) and Ludwig Sylow (1832–1918), was published in two volumes. In Abel's *Oeuvres* were a number of articles on functional equations which were published while Abel was alive. The second volume also contained the manuscript [1824?] in which he treated the functional equation[5]

$$f(\phi(x)) = f(x) + 1. \tag{2.9}$$

Although the manuscript [1824?] went unpublished during his life, Abel's teacher Brendt Holmböe (1795–1850) included [1824?] in an earlier collection of Abel's works which he published in 1839. Copies of the paper [1824?] were evidently difficult to obtain in the years prior to the publication of [Abel 1881].[6] According to Lie

[4]For more detailed information on Korkine, consult the works by Posse and Ozhigova listed in the bibliography. These works are written in Russian.

[5]Although Abel treated many functional equations, when I refer to either *Abel's functional equation* or *the* Abel equation I refer to the equation $f(\phi(x)) = f(x) + h$. The reference [1824?] is used because it is not entirely certain when Abel wrote this manuscript. In his book [1966] Azcel dates the manuscript to 1824. Lie and Sylow date it to the period prior to Abel's travels to France and Germany which commenced in 1825. Since it appears to be a follow-up to Abel's first paper on functional equations [1823], Azcel's dating does not seem unreasonable.

[6]Despite this, the Italian Carlo Formenti wrote a short paper, [1875], concerning [1824?], the Abel functional equation and another functional equation Abel considered. It appears, however,

and Sylow, copies of Holmböe's collection had become "very rare," so much so that many mathematicians, including Alfred Clebsch (1833–1872), Leopold Kronecker (1823–1891) and Weierstrass, as well as the *Société mathématique de France*, had called for a new edition of Abel's work [Abel 1881,I:i]. Their pleas resulted in the Lie-Sylow edition.

Korkine might also have been familiar with Abel's work in functional equations through Lie's discussion of Abel's work in a review of his own theory of transformation groups in *Darboux's Bulletin* in 1877. Lie pointed out that the symmetric functional equation $f(x,y) = f(y,x)$ which Abel had treated in [1826a] was a special case of a functional equation which was central to Lie's study:

$$f(f(x,a),b) = f(x,\phi(a,b)),$$

where $\phi(a,b)$ and $f(x,y)$ are both unknown [Lie 1877:383].

As will be seen shortly, the solution Abel presented [1824?] to equation (2.9) was incomplete because he reduced it to that of a difference equation without explaining how the difference equation might be solved. One can well imagine that because of the renewed appreciation of Abel's work, the possibility of remedying a problem in one of his rediscovered papers was considerably more enticing to Korkine than solving a problem posed by Schröder.

2.4 Abel's Study of Functional Equations

Abel's first paper on functional equations outlined a general method of solution involving

> two independent quantities x and y, given functions α, β, γ, δ, etc. and [unknown functions] ψ, f, F, etc.... [1823:1].

The method he gave in his paper [1823] is more of an indication of how one might go about solving a functional equation than an algorithmic procedure. Abel suggested, in essence, that one should attempt to reduce a functional equation to either a differential equation or a difference equation involving just one of the unknown functions whose solution would then yield the other unknown functions. Abel cautioned that it might not be possible to solve the reduced equation.

In his paper [1823] Abel illustrated his method by reducing several functional equations to differential equations, including

$$\phi(x) + \phi(y) = \phi(x+y) + \phi(x-y) \tag{2.10}$$

that this paper had little or no influence on either Korkine's paper or subsequent developments in the study of iteration.

which had been considered previously by Jean Le Rond D'Alembert (1717-83), Siméon-Denis Poisson (1781–1840) and Augustin-Louis Cauchy (1789–1857).[7] Abel's papers [1826a] and [1827] are illustrative of the method outlined in [1823] since the solution of the functional equations therein involved the reduction of the given functional equation to a differential equation. Only in [1824?] did Abel reduce a functional equation to a difference equation. This is significant because in [1823] he noted that time considerations prevented him from showing how the reduction to a difference equation could be used to solve a functional equation, therefore it could very well be that Abel's intention in presenting the functional equation $f(\phi(x)) = f(x) + 1$ was to provide such an example.

Abel did not, in the various papers which comprise his study of functional equations, give any explicit motivation for his interest in the subject.[8] It is quite likely that he was responding to existing studies of functional equations, but it is also possible that he saw the study of the solution of functional equations as the functional analog of his researches into the study of the solution to polynomial equations of one variable.

One mathematician to whom Abel seemed to be responding was Cauchy, who in his *Cours D'Analyse* showed that the following functional equations, sometimes called the Cauchy functional equations,

$$\begin{aligned} f(x+y) &= f(x) + f(y) \\ f(x+y) &= f(x)f(y) \\ f(xy) &= f(x) + f(y) \\ f(xy) &= f(x)f(y) \end{aligned}$$

are solved respectively by

$$\begin{aligned} f(x) &= ax \\ f(x) &= e^{ax} \\ f(x) &= a\ln(x) \\ f(x) &= x^{a}. \end{aligned}$$

That Abel was influenced by Cauchy's studies is underscored by the central role which the last of the Cauchy equations played in Abel's treatment of the binomial function $\beta(z) = (1+z)^m$ in his paper [1826b].

The posthumous fragment [1824?] consists of Abel's treatment of the functional equation

$$f(\phi(z)) = f(z) + 1 \tag{2.11}$$

[7]Both Aczel [1966] and Kuczma [1990] contain comprehensive listings of the literature of functional equations.

[8]There is no evidence suggesting that Abel was interested in the equation $f(\phi(z)) = f(z) + 1$ in order to iterate $\phi(z)$, which, the reader will recall, was the source of Schröder's interest in this equation.

where $\phi(z)$ is an arbitrary, presumably analytic, function. Although Abel does not make the fact explicit, it will be assumed that the functions involved are complex functions.

Consistent with the goals of [1823], Abel reduced the solution of the Abel equation (2.11) to a difference equation. This is done by first setting $g(z) = f^{-1}(z)$ and letting $y = f(z)$, in which case $g(y) = z$. Equation (2.11) then becomes

$$f(\phi(g(y))) = y + 1.$$

Applying $g(z)$ to both sides yields

$$\phi(g(y)) = g(y+1). \tag{2.12}$$

Abel suggested this last equation could be solved via "finite differences," but gave no indication of how to solve (2.12) in general [1824?:37]. It is likely that he expected that the solution of the difference equation would vary with the nature of $\phi(z)$. That he felt this way is indicated by his inclusion of a solution to the Abel equation in the special case $\phi(z) = z^n$, which he showed is satisfied by the function

$$f(z) = \frac{\log\log(z) - \log\log(a)}{\log(n)},$$

which he accomplished by first solving the difference equation (2.12).[9] His solution method relied on the particular properties of the function $\phi(z) = z^n$ and thus did not generalize.

As was the case with the particular functional equations Schröder solved, Abel's success in this example was due to fortuitous circumstance rather than to a general approach. Perhaps it was because he did not develop a truly general solution to the Abel equation that he left [1824?] unpublished. As Darboux observed in his review of Abel's *Oeuvres* in [Darboux 1881], Abel's unpublished works were probably left unpublished for a reason.

2.5 Korkine's Solution to the Abel Equation

The chief goal of Korkine's paper [1882] was the solution of the following version of the problem of analytic iteration. Given an analytic function $\phi(z)$, find a function $\Phi(w,z)$, analytic in the two variables w and z, (assumed to be complex although Korkine did not explicitly state this) which satisfies the following conditions:

$$\Phi(w+u,z) = \Phi(w,\Phi(u,z))$$
$$\Phi(1,z) = \phi(z).$$

[9] Abel seemed unconcerned that his solution had singularities at $z = 0$ and $z = 1$.

Korkine proposed two ways of finding the analytic iteration function $\Phi(w, z)$, one of which involved finding an invertible function $f(z)$ which satisfies the Abel functional equation

$$f(\phi(z)) = f(z) + 1. \tag{2.13}$$

The analytic iteration function would then be defined by setting

$$\Phi(w, z) = f^{-1}(f(z) + w).$$

The second method of solution involved the direct calculation of $\Phi(w, z)$.

Both of Korkine's approaches were flawed in that in each instance he assumed the existence of certain single-valued functions without providing the necessary existence proofs.

Much of Korkine's paper was devoted to his first approach wherein "the search for the function $\Phi(w, z)$ is reduced to that of the function $f(z)$," where $f(z)$ is an invertible solution to the Abel equation (2.13) [1882:233]. Before embarking on his solution of the Abel equation he remarked that

> One can find the [Abel] equation in a memoir of Abel's where he reduced its solution to that of an ordinary finite difference equation [i.e., equation (2.12)]. Therefore, since this new equation is no easier to solve than the other ... I will assay to treat the [Abel] equation directly [1882:235].

Korkine attempted to solve the Abel equation by reducing it to the solution of the functional equation

$$\Omega(\phi(z))\phi'(z) = \Omega(z), \tag{2.14}$$

a functional equation I will refer to as the Korkine equation. If an analytic solution $f(z)$ to the Abel equation exists, then its derivative $f'(z)$ is a solution to the Korkine functional equation. This can be seen by differentiating both sides of the Abel equation, which yields,

$$f'(\phi(z))\phi'(z) = f'(z).$$

This suggests that if an analytic solution $\Omega(z)$ to the Korkine equation (2.14) exists, then a properly chosen anti-derivative of $\Omega(z)$ solves the Abel functional equation. This is precisely the tack Korkine took, and there is merit to this line of reasoning. Indeed, Kuczma in [1990] reduces the solution of the Abel equation in the $\phi'(x) = 1$ case to finding an anti-derivative of a function satisfying equation (2.14). Moreover, Leau in his thesis [1897] treats the Abel functional equation in the $\phi'(x) = 1$ case by reducing it to the functional equation

$$G(\phi(z)) = \phi'(z)G(z),$$

which is satisfied by the function

$$G(z) = \frac{1}{\Omega(z)},$$

where $\Omega(z)$ is a solution of equation (2.14).

The principal defect in Korkine's approach is that his solution of the functional equation at (2.14) was by no means rigorous. It was predicated on the existence of a mysterious function $\psi(z)$ which was chosen so that the following series

$$\sum_{n=-\infty}^{+\infty} [\psi(\phi^n(z))\frac{d}{dz}\phi^n(z)],$$

converges to the function $\Omega(z)$. That this function $\Omega(z)$ *formally* satisfies the Korkine equation (2.14), that is, without regard to its convergence, is verified by direct calculation. Whether it is a convergent solution is another matter altogether, one which Korkine did not address except to say:

> As to the choice of the function $\psi(z)$, we will not occupy ourselves here, as this question depends on the nature of the function $\phi(z)$ [1882:237].

The above quotation indicates that the function $\psi(z)$ was to be chosen in an ad hoc manner, just as Abel's solution to the difference equation at (2.12) in the case $\phi(z) = z^n$ was peculiar to the nature of $z \mapsto z^n$.

Korkine's second approach to the problem of analytic iteration was to directly construct the function $\Phi(w, z)$ by first assuming that it was of the form

$$\Phi(w, z) = x + \alpha_1(w)(z - x) + \alpha_2(w)(z - x)^2 + \cdots,$$

where x is a fixed point of $\phi(z)$ satisfying $\phi'(x) \neq 0$, and the $\alpha_i(w)$ are unknown functions. Without providing the requisite existence proof that the function $\Phi(w, z)$ actually exists, he then deduced a recursive relationship among the $\alpha_i(w)$ which he used to explicitly determine them. The relationship among the $\alpha_i(z)$ is predicated on the existence of an invertible single-valued solution $f(z)$ to the Abel equation $f(\phi(z)) = f(z) + 1$ with a pole at the fixed point x.[10]

The problem with this approach, however, is that solutions to the Abel equation in the complex case are multi-valued if $|\phi'(x)| \neq 0, 1$, as Koenigs showed in his paper [1884].[11] If $\phi'(x) = 1$, Leopold Leau (1868–1940?), a student of Koenigs, stated necessary, but not sufficient, conditions that a single-valued solution of the Abel equation at (2.13) exist, namely, that $\alpha_2^2 - \alpha_3 = 0$, where α_i is the *ith* Taylor coefficient of $\phi(z)$ expanded about the fixed point x [1897:52]. Finally, also in the case where $\phi'(x) = 1$, Kuczma used a result from [Erdös-Jobitinsky 1960/61] to show that solutions of the form

$$f(z) = c_0 \log(z) + \lambda(z)$$

---

[10] This is the first instance where the existence of a fixed point entered into Korkine's discussion. However, he did not concern himself with the study of iteration near a fixed point, and there is nothing akin to Schröder's fixed point theorem in Korkine's paper.

[11] A multi-valued solution $f(z)$ is said to satisfy a functional equation involving $\phi(z)$, for example the Abel equation, if for every z such that $\phi(z)$ and z lie in the domain of definition of $f(z)$, a function element of $f(z)$ can be chosen which satisfies the functional equation around z.

occur only if $\phi(z) = (z-x)/(z+c)$, where c and c_0 are constants and $\lambda(z)$ is meromorphic on a neighborhood of x [1990:351–52].[12]

While Korkine's solution may be valid away from the fixed point x, where a single-valued function element of the solution may exist, it is certainly not valid on a neighborhood of the fixed point.

Throughout Korkine's paper are vestiges of an old style of mathematics, one which is characterized by formidable insight into the problem at hand, yet one which, due to its general lack of rigor can easily lead to mistaken conclusions, as was the case with his treatment of the Abel equation. Korkine's tendency to make unwarranted existence assumptions was duly observed by Koenigs, who in his paper [1885] remarked that

> In admitting at the onset the possibility of a solution, as well as certain hypotheses whose mutual dependence or independence remains problematic, this eminent geometer presented a solution in the form of a series [1885:386].

It will be shown in the following chapter that, in contrast to Korkine's study, Koenigs' work is characterized by a high degree of rigor.

2.6 Farkas' Solution to the Schröder Equation

In many respects, Farkas' paper serves as companion piece to Korkine's paper [1882].[13] Where Korkine approached the problem of analytic iteration through his solution of the Abel functional equation, Farkas intended to use an invertible solution $F(z)$ to the Schröder functional equation

$$F(\phi(z)) = hF(z) \tag{2.15}$$

to define the analytic iteration function $\Phi(w,z)$ by setting

$$\Phi(w,z) = F^{-1}(h^w F(z)). \tag{2.16}$$

Farkas' approach to the Schröder equation is based on a version of the fixed point theorem. Farkas' proof of his fixed point theorem is, however, not entirely satisfactory since he assumed without sufficient proof that a particular decreasing monotonic sequence converged to 0. Aside from this problem, his solution to the Schröder equation, stated below as Theorem 2.1, is rigorous, if not entirely general. It's lack of generality is a result of a number of strict hypotheses he placed on the function $\phi(z)$ which are dictated by the nature of his solution.

Despite the fact that Farkas' reliance on the Schröder equation parallels Korkine's usage of the Abel equation, Farkas showed a curious disregard for Korkine

[12] A function is meromorphic on D if it has a pole on D.

[13] Biographical information about Farkas can be found in Ortvay [1927], written in Hungarian.

and did not even mention his name. Farkas noted that Schröder's papers [1870] and [1871] were the only studies he had encountered on the "general theory of iterative functions [1884:101]." It is unlikely that he was unaware of Korkine's work since it was published in a widely circulated Parisian journal just two years previous to his own work. Perhaps he did not consider Korkine's work a "general study" of iteration since it was not based on an attracting fixed point theorem.

In any event, Farkas' interest in the problem of iteration motivated the first rigorous solution of the Schröder equation:

Theorem 2.1 (Farkas) *Let $\phi(z)$ be a complex analytic function on a disc D centered at an attracting fixed point x of $\phi(z)$, that is, a fixed point satisfying $0 < |\phi'(x)| < 1$. Suppose further that the radius of D is greater than 1. Then if*

$$\phi(z) = x + \alpha_1(z - x) + \alpha_2(z - x)^2 + \dots,$$

with

$$\sum_{i=1}^{\infty} |\alpha_i| < 1,$$

where α_i are the Taylor coefficients of $\phi(z)$ expanded about x, there exists a solution $F(z)$ on D to the Schröder functional equation

$$F(\phi(z)) = \phi'(x)F(z). \tag{2.17}$$

Although Farkas did not seem to realize it, his theorem actually implies the existence of analytic solutions to the Schröder functional equation (2.17) under the more general hypotheses that

$$\phi(x) = x \quad \text{and} \quad 0 < |\phi'(x)| < 1. \tag{2.18}$$

To see this let $\phi(z)$ be an analytic function satisfying conditions (2.18), and let

$$\psi(z) = \frac{\phi(rz)}{r},$$

where $r \ll 1$ is chosen so that on one hand

$$\sum_{i=1}^{\infty} |\alpha_i| r^{i-1} < 1,$$

while on the other, $\psi(z)$ converges on a disc of radius greater than 1 centered at x/r.

As can be easily verified, the function $\psi(z)$ has a fixed point at x/r and in addition satisfies all the hypotheses of Farkas' theorem, consequently an analytic solution $G(z)$ exists to the Schröder equation

$$G(\psi(z)) = \psi'(x)G(z).$$

By setting $F(z) = G(z/r)$, $G(z)$ can be used to define an analytic solution $F(z)$ to the corresponding Schröder equation for $\phi(z)$.

Koenigs, whose work will be discussed in the next chapter, in his paper [1884] relaxed the hypotheses of Farkas' theorem, requiring only that $\phi(z)$ be an analytic function satisfying conditions (2.18). Because his own solution of the Schröder equation utilized an entirely different approach, it is unclear whether or not he was aware that Farkas' theorem could be generalized in the manner suggested above. Koenigs, however, in remarking that

> a character which I have assayed to imprint upon my researches, either previous or current, is the reduction to a necessary minimum the number of diverse hypotheses which have served as the basis of the works of my predecessors [Koenigs 1884:s4],

gently chided Farkas for his reliance on such a restrictive set of hypotheses. His own solution to the Schröder equation on a neighborhood of a fixed point x, in contrast, required "only a single condition," namely, that $0 < |\phi'(x)| < 1$ [1884:s4].

Despite its apparent lack of generality, Farkas' solution of the Schröder equation

$$F(\phi(z)) = \phi'(x)F(z)$$

expresses a local conjugacy between $\phi(z)$ and multiplication by $\phi'(x)$, which makes rigorous an implicit assertion Schröder made during his proof of the fixed point theorem, Theorem 1.1, namely, that iteration in the neighborhood of an attracting fixed point satisfying $0 < |\phi'(x)| < 1$ acts like repeated multiplication by $\phi'(x)$.

Chapter 3

Gabriel Koenigs

3.1 Gabriel Koenigs

Gabriel Koenigs was the dominant figure in the nineteenth century study of the iteration of complex functions. Drawing on the papers of Schröder, Korkine and Farkas, Koenigs turned the study of the iteration of complex functions into a coherent and rigorously established body of work. His influence on the study of iteration continued throughout the 1890's. Not only did two of his students, Leopold Leau and Auguste Grévy (1865–1930) each examine a special case which he did not treat, but his work also stimulated papers by Paul Appell (1855–1930), Ernest Lémeray (1860-?) and others. In this chapter I treat Koenigs' own work. In the next two, I discuss the responses to his work.

Koenigs was born in Toulouse in 1858 and died in Paris in 1931. From 1879 to 1882 he studied at the École Normale Supérieure and received his doctorate from the University of Paris in 1882. He is often linked mathematically to Gaston Darboux (1842–1917). The historian Taton, in fact, referred to Koenigs as a "disciple" of Darboux [1980:446].

Darboux taught at the École Normale in Paris until 1881, and Koenigs wrote his doctoral thesis, entitled "Les propriétés infinitésimales de éspace réglé," under Darboux's direction. The relationship between Koenigs and Darboux evidently evolved to a collaborative one, and Darboux appended Koenigs' "Sur les géodésiques à intégrales quadratiques" to the fourth volume of his *Leçons sur le théorie générale des surfaces*. There is, however, nothing in Koenigs' work to suggest that Darboux influenced his interest in the iteration of complex functions. Nonetheless, Darboux's insistence that French analysis should be practiced with greater rigor than was customary at the time had a profound effect on Koenigs' approach to the study of iteration, and it is in no small measure due to his adoption of Darboux's standards of rigor that Koenigs was able to fashion the work of his predecessors into a unified theory of iteration.

After receiving his doctorate, Koenigs enjoyed a distinguished career as a mathematician. His chief fields of interest were differential geometry and applied mechanics. He lectured on mechanics at the University of Besançon from 1883 to 1885, and it was during this time that he published the three papers [1883], [1884] and [1885] which comprise his research on the iteration of complex functions. He then served as professor of mathematical analysis at the University of Toulouse for a year before returning to Paris in 1886 as an assistant professor at the École Normale.

In 1896 he was appointed professor of mechanics at the Sorbonne, after which he became increasingly drawn towards applied and experimental mechanics. Although his interest in mechanics sparked some interesting mathematics, his mathematical output waned over the years as his interest shifted towards laboratory work. Near the beginning of World War I, he founded a laboratory which played an important role in the development of French military technology. After the war, he received a commendation from the French government in recognition of his contributions to the military effort.

Over the course of his career, Koenigs was awarded several prizes by the French Academy of Sciences, and in 1892, along with Paul Appell and several others, he was nominated for the vacant seat in the Academy's mathematics section occasioned by the death of Ossian Bonnet (1819–1892). The seat was awarded to Appell, but in 1918 Koenigs was honored with membership in the mechanics section of the Academy.

3.2 Koenigs and Darboux

Darboux's interest in the foundations of analysis represents something of a departure from his principal concern, the study of differential geometry, and was evidently motivated by his conviction that French analysis was not practiced with sufficient rigor. His study of foundational issues, which consists of the three papers [1872], [1875] and [1879], is characterized by an artful blend of precise, carefully argued proofs and insightful counter examples.

Essential to Darboux's approach to analysis is the concept of uniform convergence. In his paper [1875] he established several important facts regarding the uniform convergence of a series of single variable real-valued functions, including the theorem which asserts that the sum of a series of uniformly convergent continuous functions is itself continuous. Uniform convergence also animated many of the counter examples Darboux provided. For instance, one of the most interesting examples from [1875] is that of a continuous real function $g(x)$ which is nowhere differentiable. Darboux proved the continuity of $g(x)$ by showing that it is the sum of a uniformly convergent series of continuous functions.

In comparison with the situation in Germany and Italy, the French were late in developing a rigorous approach to analysis. Consequently, Darboux's contributions in this direction were largely ignored for quite some time. According to the historian Gispert, it was not until the latter half of the 1880's that Darboux's approach to analysis began to take hold in France [1983:63].

Koenigs was evidently one of the first French mathematicians to respond favorably to Darboux's new approach, and in his paper [1884] he adapted two of Darboux's theorems on the uniform convergence of real functions to the study of complex functions. Concomitant with his incorporation of Darboux's innovations came an increased attention to rigor. Koenigs' first work on iteration, his paper [1883], is characterized by a certain vagueness and lacks the precision and clarity of his paper [1884]. There is considerable overlap between these two papers, and virtually without exception ideas which are discussed in both papers are subject to a more rigorous treatment in the 1884 paper.

That Darboux was responsible for this change in approach is indicated by the following quotation from Koenigs' introduction to his paper [1884]:

> The nature of the topic demands the use of the most general theorems from the theory of functions. I was principally inspired by the excellent memoir *Sur les fonctions discontinues* which Darboux published in Volume IV of the second series of *Annales de l'École Normale.*
>
> An easy extension of the results from this memoir to complex quantities ... yields the following theorems, which serve as the base of my work [1884:s4].

The two theorems which Koenigs then listed can be summarized as follows.

Theorem 3.1 (Koenigs-Darboux) *Let the functions $u_i(z)$ be analytic in a region D. Then, if the infinite series $\sum u_i(z)$ is uniformly convergent in D, its limit function $u(z)$ is continuous on D. If, in addition, $\sum u_i'(z)$ converges uniformly in D then it converges to $u'(z)$, and $u(z)$ is thus analytic in D.*

As Koenigs himself indicated, his theorems are routine extensions of those Darboux proved for real functions in [1875].

Although Koenigs did not seem to realize it, the application of the Cauchy integral formula leads to the stronger result that if a series of analytic functions $\sum u_i(z)$ converges uniformly on D to $u(z)$, then $u(z)$ is analytic on D.[1]

[1] Weierstrass first published a proof of this theorem in his paper [1880]. Weierstrass proved it initially, however, in the manuscript [1841] which went unpublished until the 1890's. That neither Koenigs nor Appell, who used Koenigs' two theorems in his paper [1891:285], realized that these theorems could be condensed into one is indicative of the lack of communication between French and German mathematicians. It may also reflect a lack of emphasis on complex function theory within the French mathematical community of the time.

3.3 The Background to Koenigs' Study of Iteration

In his introductions to both [1884] and [1885] Koenigs discussed his work in the context of his predecessors. He noted that he grounded his work on a fixed point theorem in spirit equivalent to Schröder's fixed point theorem, Theorem 1.1. However, in contrast to Schröder who never applied Theorem 1.1 to the study of functional equations, Koenigs used his own fixed point theorem to rigorously treat both the Schröder equation

$$F(\phi(z)) = \phi'(x)F(z)$$

and the Abel equation

$$f(\phi(z)) = f(z) + 1$$

in a neighborhood of a fixed point x satisfying $0 < |\phi'(x)| < 1$. Moreover, he revealed a connection between the Abel and Schröder equations, to be discussed below, that had previously escaped notice.

In his introductory remarks Koenigs also noted that his approach repaired defects in the work of both Korkine and Farkas. As was discussed in the previous chapter Koenigs criticized Korkine's treatment of the Abel equation because it is grounded on unsubstantiated, and in general incorrect, existence claims, and also observed that the hypotheses Farkas required for his solution to the Schröder equation are not of sufficient generality.

That Koenigs' interest in iteration evidently grew out of strictly mathematical concerns rather than physical ones is interesting, since the folk history of mathematics often has it that complex dynamics grew out of Henri Poincaré's (1854–1912) study of celestial mechanics. While it is true that Poincaré iterated real-valued solutions of certain differential equations as early as his paper [1881], a number of factors augur against his being a principal influence on Koenigs, the most salient of which is that although Koenigs referenced Poincaré's work involving linear fractional transformations, nowhere did he cite Poincaré's studies of mechanics.[2]

Given Poincaré's prominence by the mid 1880's, and given Koenigs' proclivity to discuss the work of his predecessors, it seems likely that, had he thought an important connection existed with Poincaré's use of iteration, Koenigs would have been eager to point it out. Moreover, Koenigs' use of iteration was quite different from that of Poincaré. Not only did Koenigs iterate complex analytic functions while Poincaré iterated real functions, but in contrast to Koenigs' highly formalized study of iteration, Poincaré's early use of iteration was relatively unstructured.

Perhaps it is the case that Poincaré's work in celestial mechanics did influence Koenigs' study. There is, however, nothing in Koenigs' work which suggests this. Nor is there anything in the works of Fatou and Julia which suggests that Poincaré's

[2] In his paper [1885], Koenigs noted that Poincaré had shown the existence of a family of circles passing through the fixed points x, x' of a linear fractional transformation $L(z)$ which were invariant under iteration [1885:404].

study of mechanics influenced their own studies of iteration. When they do refer to those mathematicians who paved the way for their studies, it is Koenigs' name which figures most prominently.

3.4 Koenigs' Study of Fixed Points

The fundamental object of Koenigs' study is the set of accumulation points of the sequence

$$\{z_0, \phi(z_0), \phi^2(z_0), \ldots\}, \tag{3.1}$$

where z_0 is fixed, and $\phi(z)$ is an analytic function defined on a domain G, that is, a connected open subset G of the extended plane $\bar{\mathbb{C}}$. The sequence at (3.1) is nowadays called the *forward orbit of* z_0.[3]

Koenigs focused on two cases, the first, where sequence (3.1) has a unique limit point x, and, the second, where the set of the accumulation points of (3.1) is the finite set

$$\{x_0, x_1, \ldots, x_{p-1}\}.$$

If sequence (3.1) has a unique limit point x, then $\phi(x) = x$ since the continuity of $\phi(z)$ in conjunction with the limit

$$\lim_{n\to\infty} \phi^n(z_0) = x$$

implies that

$$\phi(x) = \phi(\lim_{n\to\infty} \phi^n(z_0)) = \lim_{n\to\infty} \phi^{n+1}(z_0) = x.$$

In the second case it can be shown that

$$\lim_{n\to\infty} \phi^{np}(z_0)$$

equals one of the x_i, and that $\phi(x_i) = x_{i+1}$, with $x_p = x_0$.

The point x is called a *fixed point* and the points x_i are called *periodic points of period* p, or, more succinctly, *period* p points. The formal definition of a period p point is as follows:

Definition 3.2 *A point x_i is a* periodic point of period p, $p \geq 1$, *if p is the smallest positive integer such that $\phi^p(x_i) = x_i$. Such a point is often referred to as a* period p point. *A fixed point is thus a period* 1 *point. A periodic* orbit *is the set $\{x_0, \ldots, x_{p-1}\}$, where $x_i = \phi^i(x_0)$ for $0 \leq i \leq p$ and $x_p = x_0$.*

[3]The *backward orbit* of z_0 under $\phi(z)$ is the set

$$\{z, \phi^{-1}(z_0), \phi^{-2}(z_0), \ldots\},$$

where $\phi^{-1}(z)$ is, depending on context, either the total inverse of $\phi(z)$ or a fixed inverse.

Since period p points of $\phi(z)$ are fixed points of $\phi^p(z)$, any theorem regarding a fixed point of $\phi(z)$ also applies to periodic points via the function $\phi^p(z)$, and Koenigs therefore reduced the study of periodic points of $\phi(z)$ to the study of the fixed points of $\phi^p(z)$. Although Schröder examined periodic points in specific instances, for example, in his investigation of the Newton's method function for the quadratic, they played a significant role neither in his work nor in that of Farkas or Korkine, and Koenigs was the first to treat such points systematically.

As has been noted, one of the characteristics of Koenigs' study is its rigor. This is especially the case with the definition from his paper [1884] of the convergence of a sequence $\{\alpha_n\}$ towards a limit x. This definition is presented far more precisely than anything found in the papers of his predecessors—or, as the following quotation suggests, anything from his earlier paper:

> I recall in terms more general than I used in my preceding memoir, and in a more complete and precise fashion [the following]: The sequence $\alpha, \alpha_2, \alpha_3, \ldots, \alpha_n, \ldots$, is said to converge regularly towards a limit x when for all positive ϵ, as small as one wishes, it is possible to find a number N_ϵ large enough so that under the *sole* condition that $n \geq N_\epsilon$, one has $|\alpha_n - x| < \epsilon$ [1884:s5–6].[4]

The reason that Koenigs was interested in the convergence of $\{|\alpha_n - x|\}$ is that, as Schröder asserted in his paper [1870], and Koenigs verified in [1884], the convergence under iteration of $\phi(z)$ to a fixed point x satisfying $|\phi'(x)| < 1$ is such that there exists a disc D centered at x such that the sequence

$$\{\phi^n(z) - x\}$$

converges to 0. Because of this, the fixed point x is usually referred to as an attracting fixed point.

Koenigs built his entire study of iteration on the case where sequence (3.1) converges to a periodic p orbit, where $p \geq 1$. It may seem a little surprising that Koenigs limited himself to this particular case. However, it can be shown that if the set of accumulation points to this sequence is finite, then the accumulation points are period p points. It will be seen in later chapters that in general the above sequence has only finitely many accumulation points for all z in $\bar{\mathbb{C}} - J$, where J is a certain subset of $\bar{\mathbb{C}}$ called the Julia set. The case where $z \in J$, that is, where the sequence (3.1) has infinitely many accumulation points is very difficult, and the first systematic study of this last case did not occur until Fatou's note [1906a].

If the sequence

$$\{z, \phi(z), \phi^2(z), \ldots\}$$

[4]Koenigs' use of the term regular was not intended to suggest monotone convergence, but rather meant that the α_n had a unique limit. In [1883] he also spoke of irregular convergence, by which he meant that the α_n had more than one subsequential limit, as was the case when $\phi^n(z_0)$ converges to a periodic orbit. He later dropped irregular convergence in favor of the term periodic convergence [1884:s6].

is the fundamental object in Koenigs' study, then the following theorem, Koenigs' version of the fixed point theorem first articulated by Schröder in his paper [1870], is his fundamental theorem:

Theorem 3.3 (Koenigs' Fixed Point Theorem) *If $\phi(z)$ is analytic on a domain G contained in $\bar{\mathbb{C}}$ which contains a fixed point x of $\phi(z)$ satisfying*

$$|\phi'(x)| < 1,$$

then there exists a disc D surrounding x such that

$$\lim_{n\to\infty} \phi^n(z) = x$$

for all z in D.

Koenigs also proved the following partial converse:

Theorem 3.4 (Koenigs) *If on the other hand, the sequence $\{\phi^n(z)\}$ remains interior to a domain D for all $z \in D$, and in addition converges to a fixed point x which is not necessarily in D, then $|\phi'(x)| \leq 1$.*

The partial converse implies that the fixed point x must be on the boundary of D if it is not in D, although it should be noted that Koenigs did not use any set theoretic terms other than domain, by which he meant what we would call an open, connected set.

Koenigs gave two proofs of his fixed point theorem, one in [1883] and another in [1884]. He proved the partial converse only in [1883], merely restating it in [1884], although the equal part of the inequality $|\phi'(x)| \leq 1$ was inadvertently left out, probably due to a typographical error [1884:s6]. The reason that Koenigs proved his fixed point theorem again in [1884] was evidently to repair a defect in his earlier proof.

Koenigs began the proof in [1883] by correctly showing that if $\phi(x) = x$ and $|\phi'(x)| < 1$, then there exists an open disc D of radius r centered at the fixed point x such that for all z in D,

$$\frac{|\phi(z) - x|}{|z - x|} < 1. \tag{3.2}$$

This implies that for all z in D, $\phi(z)$ is closer to x than z , since $|\phi(z) - x| < |z - x|$. Substituting $\phi(z)$ for z in the inequality at (3.2) implies in turn that $\phi^2(z)$ is closer to x than $\phi(z)$, since

$$\frac{|\phi(\phi(z)) - x|}{|\phi(z) - x|} < 1,$$

and in general, that $\phi^n(z)$ is closer to the fixed point x than is $\phi^{n-1}(z)$. Koenigs evidently disregarded the possibility that $\phi^n(z)$ could continually get closer to x yet remain bounded away from x and ended his proof with a brief remark to the effect that since the $\phi^n(z)$ were getting closer to x "without ceasing," the $\phi^n(z)$ "had x for its limit [1883:345]."

In his paper [1884], he gave a stronger inequality than the one at (3.2), namely, that for all z in D,

$$\frac{|\phi(z) - x|}{|z - x|} < \lambda < 1. \tag{3.3}$$

This observation gave him the means to rigorously prove Theorem 3.3. I will outline his argument, since it typifies a sort of convergence proof often used in the study of complex dynamics.

Since $|\phi'(x)|$, D can be chosen so that

$$\frac{|\phi(z) - x|}{|z - x|} < \lambda < 1$$

on D, that is,

$$|\phi(z) - x| < \lambda|z - x|.$$

Substituting $\phi(z)$ for z in this last inequality implies that

$$|\phi^2(z) - x| < \lambda|\phi(z) - x| < \lambda^2|z - x|.$$

Hence, continually substituting $\phi(z)$ for z yields

$$|\phi^n(z) - x| < \lambda^n|z - x|. \tag{3.4}$$

Since $\lambda < 1$ the convergence of $\phi^n(z)$ to x is immediate.

Koenigs' proof also shows that the convergence under iteration to x is uniform on D: since all z in D satisfy $|z - x| < r$, where r is the radius of D, inequality (3.4) implies that

$$|\phi^n(z) - x| < \lambda^n|z - x| < r\lambda^n$$

for all z in D. Thus, for any given $\epsilon > 0$, n can be chosen so that for all $z \in D$, $|\phi^n(z) - x| < \epsilon$. The increased attention to detail seen in the proof from [1884] is evident throughout [1884], and it attests to the influence of Darboux.

Koenigs extended Theorem 3.3 to the case where x is a period p point, that is, where p is the smallest positive integer such that $\phi^p(x) = x$. As noted above, he reduced the study of period p points to that of fixed points because a period p point of a function $\phi(z)$ is a fixed point of $\phi^p(z)$. An attracting period p point x of $\phi(z)$ is therefore one which satisfies $|\frac{d}{dz}\phi^p(x)| < 1$. According to Theorem 3.3, for each point x_i of the form $x_i = \phi^i(x_0)$ (with $x_0 = x_p$), there exists a disc D_i centered at x_i such that the function $\phi^i(z)$ converges under iteration to x_i for all z on D_i. Finally, since repeated application of the chain rule implies that

$$M = \frac{d\phi^p}{dz}(x_0) = \phi'(x_0)\phi'(x_1)\cdots\phi'(x_{p-1}),$$

the quantity M is independent of the choice of x_i.

This reduction of a periodic p point to a fixed point leads to the following definition, which is based on one given by Koenigs in [1883].

Definition 3.5 *A point* x_i *from a periodic orbit* $\{\phi^i(z)\}$ *is an* attracting period p point *if* $|\frac{d\phi^p}{dz}(x_i)| < 1$. *The orbit* P *is called an* attracting periodic orbit *or* an attracting period p orbit *since, by Theorem 3.3, for each* x_i *there exists a disc* D_i *centered at* x_i *such that the function* $\phi^p(z)$ *converges under iteration to* x_i *for all points on* D_i. *Therefore given any point* z *in* D_i *the accumulation points of the forward orbit of* z *consist of the periodic orbit* $\{x_i\}$.

Koenigs also realized that the Riemann sphere $\bar{\mathbb{C}}$ is the natural place to study iteration of complex functions, and consequently extended his study so as to allow for the possibility that the point at ∞ may be a fixed point. For example, for any polynomial $\phi(z)$, the point at ∞ acts just like an attracting fixed point outside a sufficiently large neighborhood of the origin. To see this, consider the special case $\phi(z) = z^2$. Since $\phi^n(z) = z^{2^n}$, any point z exterior to the unit disc is attracted to ∞ in the sense that as n approaches ∞ so does $\phi^n(z)$.

To make the study of iteration near a fixed point at infinity rigorous, Koenigs used the coordinate change $z \mapsto 1/z$, which maps the fixed point at infinity to the origin, and which has since become the standard way to treat fixed points at infinity.

Using this coordinate change, the derivative at infinity is

$$\phi'(\infty) = \frac{1}{\psi'(0)},$$

where

$$\psi(z) = \frac{1}{\phi(1/z)},$$

which is analytic in a neighborhood of the origin. For a polynomial $\phi(z) = a_n z^n + \cdots + a_0$, the map $\psi(z)$ is

$$z \mapsto \frac{z^n}{a_0 z^n + \cdots + a_n}.$$

The relationship between $\phi(z)$ and $\psi(z)$ is given in the following diagram:

$$\begin{array}{ccc} D^* & \xrightarrow{\quad\psi\quad} & D^* \\ \Big\downarrow \frac{1}{z} & & \Big\downarrow \frac{1}{z} \\ D & \xrightarrow{\quad\phi\quad} & D \end{array}$$

where D is a neighborhood of infinity, that is, the set $\{z : |z| > r\}$ for some sufficiently large r, and D^* is the corresponding neighborhood of the origin $\{z : |z| < 1/r\}$. Just as in the case where x is finite, under this coordinate change ∞ is an attracting fixed point if $|\phi'(\infty)| < 1$. Koenigs' comments on the matter are interesting, if only as an indication of what he thought might be shocking to a nineteenth century mathematician:

> The expression of regular convergence towards infinity is thus explained easily by the preceding, and it is not anything very shocking if one makes use of the sphere for representation [1883:351].

That Koenigs' careful treatment of the fixed point at infinity is contained in [1883] is noteworthy since it indicates that not everything in [1883] was treated heuristically.

The extension of the study of iteration to the Riemann sphere also accounts for the emphasis on rational maps seen in the work of Fatou and Julia since rational maps are the only functions which are analytic on the entire sphere.

3.5 Koenigs' Solution of the Schröder Equation

As noted at the conclusion of the previous chapter, Farkas proved that if an analytic function $\phi(z)$ has an attracting fixed point x such that $0 < |\phi'(x)| < 1$, and, in addition, satisfies some rather strict hypotheses (see Theorem 2.1), then an analytic solution $B(z)$ to the following Schröder functional equation exists on a neighborhood of x:

$$B(\phi(z)) = \phi'(x)B(z). \tag{3.5}$$

One of Koenigs' major accomplishments was to simplify Farkas' hypotheses considerably and show that a sufficient condition for the existence of an analytic solution to (3.5) on a neighborhood of the fixed point x is that $0 < |\phi'(x)| < 1$. He deduced from this that a solution also exists if $|\phi'(x)| > 1$.

Koenigs found the study of the Schröder equation in particular and iteration in general very difficult when $\phi'(x)$ is zero or one in modulus. Consequently, he produced no results in either of these cases. The study of these two cases, however, was taken up in the 1890's by Koenigs' students. His student Leau, and another French mathematician, Ernest Lémeray (1860–?) each produced some preliminary results in the case where $\phi'(x)$ is a root of unity. Grévy, another of Koenigs' students, studied several generalized versions of the equation (3.5) in the case where $\phi'(x) = 0$.

In what follows, I will adopt a convention to which I will adhere in the sequel, namely, I will take the fixed point x to be 0, which can be accomplished via a change of coordinates. The Schröder equation at (3.5) will be expressed as

$$B(\phi(z)) = \phi'(0)B(z), \tag{3.6}$$

and will be referred to as the canonical Schröder equation.

Koenigs' solution to the canonical Schröder equation is defined as follows:

$$B(z) = \lim_{n\to\infty} \frac{\phi^n(z)}{(\phi'(0))^n}. \tag{3.7}$$

That $B(z)$ formally satisfies the Schröder equation at (3.6) is seen by direct calculation:

$$\begin{aligned} B(\phi(z)) &= \lim_{n\to\infty} \frac{\phi^n(\phi(z))}{(\phi'(0))^n} \\ &= \lim_{n\to\infty} \frac{\phi^{n+1}(z)}{(\phi'(0))^{n+1}}\phi'(0) \\ &= \phi'(0)B(z). \end{aligned}$$

Koenigs in his paper [1884] demonstrated the analyticity of $B(z)$ by reducing the convergence of the limit (3.7) on a neighborhood D of the fixed point to that of a certain series of functions, $\sum \beta_i(z)$. He then proved that both series $\sum \beta_i(z)$ and $\sum \beta_i'(z)$ converge uniformly on D, and deduced the analyticity of $\sum \beta_i(z)$ from the theorem he borrowed from Darboux, Theorem 3.1 above. This theorem asserts that if a series of functions $\sum u_i(z)$ converges uniformly on D, and if $\sum u_i'(z)$ does as well, then the series $\sum u_i(z)$ converges to an analytic function $G(z)$ on D. Finally, Koenigs demonstrated that the function $B(z)$ also satisfies $B(0) = 0$ and $B'(0) = 1$, hence, the function $B(z)$ is invertible on D.[5]

Koenigs' treatment of the Schröder equation serves as another example of the increased precision which accompanied his second work, [1884]. Koenigs' first paper, [1883], featured a somewhat imprecise proof that the limit converges pointwise on D. Although he did not explicitly say that this pointwise convergence implied that the limit was analytic, the feeling lingers that he thought this was so.

3.6 Koenigs and Functional Equations

The most extensive use to which he put the theory which he developed was the application of his solution $B(z)$ of the canonical Schröder equation

$$B(\phi(z)) = \phi'(0)B(z), \tag{3.8}$$

to the solution of other functional equations. As Koenigs noted in his introduction to [1884]:

[5]Koenigs' proof from [1884] is quite long (see [1884:s7–s16]). A short proof of the convergence can be found in Chapter 6 of Milnor's preprint [1990].

> There exist, moreover, infinitely many functional equations to which my method extends and to which the function $B(z)$ yields a general solution [1884:s4].

Koenigs in fact devoted the latter portion of [1884] and virtually all of [1885] to this pursuit.

The first functional equation to which he applied his method, was the general Schröder equation

$$F(\phi(z)) = hF(z), \tag{3.9}$$

which, in the event that $h = (\phi'(0))^k$, for integer k, is solved in the neighborhood of a fixed point 0 satisfying $0 < |\phi'(0)| < 1$ by the function

$$F(z) = c(B(z))^k, \tag{3.10}$$

where c is arbitrary, as can be verified by direct calculation. Koenigs showed as well that the general Schröder equation at (3.9) has an analytic or meromorphic solution near x only if $h = (\phi'(0))^k$.

In his paper [1885], he reduced the solution of the Schröder equation

$$F(\phi(z)) = \phi'(0)F(z),$$

where 0 is a repelling fixed point, that is, $1 < |\phi'(0)| < \infty$, to the solution of the canonical Schröder equation at (3.8). In this event, let $\psi(z)$ be the local inverse of $\phi(z)$ which satisfies $\psi(0) = 0$. Since $\psi'(0) = 1/\phi'(0)$ and is therefore strictly between zero and one in modulus, there exists a locally defined function $B(z)$ which satisfies the canonical Schröder equation

$$B(\psi(z)) = \frac{1}{\phi'(0)}B(z).$$

Letting $z \mapsto \phi(z)$, and multiplying both sides of the above equation by $\phi'(0)$ yields

$$B(\phi(z)) = \phi'(0)B(z).$$

Although Koenigs did not do so, an equivalent way of treating the case where $|\phi'(0)| > 1$ is to study the equation

$$F(\phi'(0)z) = \phi(F(z)).$$

This method was favored by both Samuel Lattès and Joseph Fels Ritt (1893–1951) in their respective studies of iteration circa 1918, which are discussed in Chapter 10.

Koenigs also discovered an important link between the function $B(z)$ and the Abel equation

$$b(\phi(z)) = b(z) + 1$$

on a neighborhood D of 0, where $0 < |\phi'(0)| < 1$. Let $\log(z)$ denote an appropriate branch of the logarithm and define $b(z)$ formally as follows:

$$b(z) = \frac{\log(B(z))}{\log(\phi'(0))}, \tag{3.11}$$

where $B(z)$ is the solution of the canonical functional equation for $\phi(z)$. Then

$$\begin{aligned} b(\phi(z)) &= \frac{\log B(\phi(z))}{\log \phi'(0)} \\ &= \frac{\log B(\phi'(0)B(z))}{\log \phi'(0)} \\ &= \frac{\log B(z) + \log \phi'(0)}{\log \phi'(0)} \\ &= b(z) + 1. \end{aligned}$$

Since $B(0) = 0$, $b(z)$ is multi-valued near the origin and can only be thought of as a solution in the sense that it is possible to find a single-valued locally analytic function which satisfies the Abel equation on simply connected subdomains D^* of D which do not contain the origin, provided that care is taken in choosing D^* so that equality

$$\log(AB) = \log(A) + \log(B)$$

holds. Koenigs, in addition, pointed out that $b(z)$ is the best possible solution to the Abel equation since there are no analytic or meromorphic solutions to it in a neighborhood of an attracting fixed point 0 satisfying $0 < |\phi'(0)| < 1$.

The major concern of the paper [1885] was Koenigs' application of $B(z)$ to various problems including analytic iteration and the solution of new functional equations, among them,

$$\begin{aligned} F^p(z) &= \phi(z), \\ F(\phi(z)) &= \phi(F(z)), \\ F(\phi(z)) &= \psi(F(z)) \end{aligned} \tag{3.12}$$

where $\phi(z)$ and $\psi(z)$ are given analytic functions.

In the next section I will discuss the principal shortcoming of Koenigs' study, his failure to develop a global study of iteration. However, it is important not to lose sight of the fact that Koenigs' study was very successful. He discovered heretofore unknown connections between the various strands of his predecessors' work— in particular Schröder's fixed point theorems and the study of the Abel and Schröder equations—and established a unified, rigorous local theory regarding the iteration of complex analytic functions which he then applied to the study of various functional equations. As will be seen in subsequent chapters, Koenigs influenced the study for quite some time.

3.7 Koenigs and the Global Study of Iteration

Koenigs based his entire study of iteration, as well as his related study of functional equations, on his fixed point theorem, Theorem 3.3, which states that if $\phi(0) = 0$ and $|\phi'(0)| < 1$, then there exists an open disc D centered at 0 on which $\phi(z)$ converges under iteration to 0 for all z in D. Due to the fact that his fixed point theorem holds only on the disc D, the scope of Koenigs' study is local rather than global since his results say nothing about the behavior of the iterates of an arbitrary point z in $\bar{\mathbb{C}}$ under a given function $\phi(z)$.

Koenigs, not surprisingly, was clearly not satisfied with the local nature of his theory of iteration, and at the end of his paper [1884] he suggested a way to extend the study of iteration beyond the neighborhood D. His method in essence is as follows.

Let D be the disc surrounding 0 on which all points converge under iteration by $\phi(z)$. Choose this disc so that the solution $B(z)$ to the canonical Schröder equation exists on D. Suppose as well that there are points which converge to 0 under iteration by $\phi(z)$ but are not in D, and let $\tilde{z}$ be one such point. There will then be a small disc $\tilde{D}$ surrounding $\tilde{z}$ on which points converge to 0 under iteration and to which his theorems could be extended [1884:s40].

Although Koenigs did not explicitly prove the existence of such points as $\tilde{z}$, it is not unreasonable to suppose they exist since the function $\phi(z)$ is generally a many-to-one mapping. However, he used the fact that $B'(0) = 1 \neq 0$ to define D so that $B(z)$ is one-to-one on D, which implies that $\phi(z)$ is also one-to-one on D. Thus, in the event that $\phi(z)$ is not globally one-to-one, there are points in the preimage of D under $\phi(z)$ which are not in D.

The following quotation indicates that Koenigs understood the problems involved in extending his study of iteration beyond the neighborhood of a fixed point.[6]

> If one envisions all the points which [iterate] to the interior of D and consequently [converge] to the point 0, one can extend general theorems to this region. But one knows nothing of the general manner in which this region is limited, and one cannot affirm *a priori* that the mode of delimitation is not of a nature which restricts this extension [1884:s40].

The pessimism expressed in the above quotation stems from another problem which seems to have frustrated Koenigs, namely, to partition $\bar{\mathbb{C}}$ into regions A_k according to the behavior of the iterates of points in A_k. For example, given an attracting fixed point 0, or any attracting periodic orbit $\{x_i\}$, one such A_k would be the entire set of points which converge to 0, or to the orbit $\{x_i\}$.[7] He expressed this desire explicitly in his earliest paper (see [1883:35]), and referred to it again in the following quotation:

[6] As is the case with many of the quotations used, notation has been adjusted to conform with my own notational conventions.

[7] For the definition of an attracting periodic orbit, see Definition 3.5 above.

> The importance of the division of the plane into regions according to the limit points to which these points [converge under iteration] has thus come to the fore. But the difficulty attached to the problem is evident when one realizes that there are an infinity of circular limit groups [that is, period p points], since the index [that is, p] to which they belong is arbitrarily large.
>
> Cayley was the first to pose this problem, in the case of Newton's method; but even in the case of a simple entire polynomial, the number of these limit groups can be infinite, and even though the problem has been solved for the equation of the second degree, the [period p points] are not limited in this case, and they all lie on the line made up of points equidistant from the root points, a line which, as one knows, divides the plane into two regions such that all points on one side of the plane are led to the limit point on that side [1884:s40–41].

The above quotation indicates that Koenigs was troubled by the fact that as n goes to infinity so does the number of solutions to $\phi^n(z) = z$. This in turn suggested to Koenigs that not only were there in general infinitely many periodic points, but that there may well be infinitely many attractive periodic orbits, hence division of the extended plane into the regions A_k would be quite complicated, and perhaps even impossible. The reference to Cayley in the quotation above is rather telling since, in characterizing the behavior of all points in $\bar{\mathbb{C}}$ according to whether or not they converge to a root of the quadratic under iteration by Newton's method, Cayley did solve the problem of "the division of the plane," and thereby set a model of perfection which Koenigs was unable to duplicate. Koenigs' remarks regarding Newton's method are interesting as well in their neglect of Schröder.

Koenigs' belief that the existence of infinitely many periodic points precluded an easy understanding of the global behavior of iteration is right on the mark, since, as will be shown in the sequel, the key to such an understanding lies precisely in sorting out the entire set of periodic points.

Koenigs' failure to understand this set, however, was more a function of history than a failure of mathematics since the mathematical tools he needed were not extant in his time. In order to achieve a thorough understanding of the set of all periodic points an understanding of set theory—in particular totally disconnected perfect sets—is needed, and, as the studies of Fatou and Julia indicate, an understanding of Montel's theory of normal families of complex functions is particularly helpful.

At the time of Koenigs' papers on iteration the rigorous study of sets was in its infancy, and examples of totally disconnected perfect sets were only just beginning to circulate within the mathematical community. Moreover, Montel's first results on normal families would not be published until the early years of the next century.

Although Fatou published a paper concerning the global solution of Schröder's equation for a limited class of functions in his paper [1906a], it wasn't until 1917 that Fatou and Julia began to establish general results concerning the global behavior of the iterates of arbitrary rational functions.

Chapter 4

Iteration in the 1890's: Grévy

4.1 A Brief Survey of Iteration in the 1890's

Although Koenigs contributed no new works to the study of iteration of complex functions during the 1890's, he nonetheless remained its central figure. The infusion of new mathematical ideas—in particular Montel's theory of normal families—which would prove so useful to the studies of Fatou and Julia did not occur until after the turn of the century. Consequently, the developments of the nineties consisted largely of either the application of Koenigs' theory into other branches of mathematics or in the extension of Koenigs' local study into two special cases he did not examine, namely, the case where the derivative at the fixed point $x = 0$ is 0 or 1 in modulus.

Two of Koenigs' students, Leopold Leau and Auguste-Clémente Grévy, each investigated one of these special cases. Grévy investigated the $\phi'(0) = 0$ case, often called the *superattracting* case, in his doctoral thesis [1894], and Leau in his dissertation [1897] studied the $|\phi'(0)| = 1$ case, often referred to as the *neutral* case.

The study of the neutral case actually began with Ernest Lémeray, who treated the $|\phi'(0)| = 1$ case in several short articles in the latter half of the 1890's. He and Leau each produced theorems which anticipate the so-called Flower Theorem (see Theorem 5.1 below). Like Leau and Grévy, Lémeray was influenced by Koenigs, a fact he readily acknowledged in the opening paragraphs of his first paper on the subject [1895].

Carlo Bourlet (1866–1913) also wrote a few articles concerning the iteration of complex functions towards the end of the 1890's, among them his papers [1899a] and [1899b]. Although his works were important to the development of the study of functional analysis, his works made no significant contributions to the march of ideas which led to the work of Fatou and Julia. In Bourlet's work, as in most of the works cited above, the study of functional equations went hand in hand with the study of the iteration of complex functions, just as it did in the 1880's.

Applications of Koenigs' work in other areas of mathematics while not plentiful were by no means non-existent. Notable were Bourlet's study of iteration as a functional operator in his paper [1899a] and Paul Appell's paper [1891] in which he used a function defined by Koenigs to solve a special case of the Hill differential equation

$$\frac{d^2u(z)}{dz^2} - u(z)f(z) = 0. \tag{4.1}$$

Although Koenigs confined himself to the iteration of single variable complex functions, several late nineteenth and early twentieth century mathematicians including Leau, the Polish mathematician Lucyan Böttcher, the American Albert Bennett and the French mathematician Samuel Lattès drew on Koenigs' research in their studies of iteration of functions of more than one variable (see Chapter VI and VII of [Leau 1897], [Böttcher 1897], [Lattès 1907] and [1908], and [Bennett 1915a]). Since the purpose of this work is to trace the development of the body of study which led to the works of Fatou and Julia concerning the iteration of complex functions of a single variable, the iteration of multi-variable functions will not be discussed.

4.2 Appell's Application of Koenigs' Work to Hill's Differential Equation

The Hill equation was considered by the American mathematician George William Hill (1838–1914) in the 1870's in connection with the orbit of the moon. Hill stipulated that $f(z)$ be periodic and sought periodic solutions $u(z)$. Although Appell made no such restriction on either $u(z)$ or $f(z)$ in his paper [1891], he required that $f(z)$ satisfy a particular functional equation, stated below. He made no mention of the relation of the Hill equation to celestial mechanics in [1891] and instead treated it as a purely mathematical object.

Appell's work [1891] is noteworthy not only because it applies Koenigs' ideas to the study of differential equations, but also because it demonstrates that Appell had an active interest in the study of iteration. This is of import because Appell was a member of the commission of the French Academy of Sciences charged with judging the 1918 *Grand Prix des Sciences mathématiques*, which was to be awarded to the best paper the Academy received regarding the iteration of complex functions.

Appell relied extensively on Koenigs' [1884] for his solution to the Hill differential equation. In fact, he wrote his particular solution to the Hill equation entirely in terms of the function $B(z)$, where $B(z)$ is Koenigs' solution to the canonical

Schröder equation

$$B(\phi(z)) = \phi'(0)B(z), \tag{4.2}$$

where $\phi(z)$ satisfies the fixed point conditions

$$\phi(0) = 0 \quad \text{and} \quad 0 < |\phi'(0)| < 1. \tag{4.3}$$

Consistent with its dependence on Koenigs' fixed point theorem, the function $B(z)$ is defined only on a neighborhood of the fixed point 0 (see Theorem 3.3).

Appell obtained the following fundamental set of solutions to (4.1), namely, the functions

$$\frac{1}{\sqrt{B'(z)}} \quad \text{and} \quad \frac{B(z)}{\sqrt{B'(z)}},$$

provided that the function $f(z)$ from the statement of the Hill equation at (4.1) satisfies the functional equation

$$f(\phi(z)) = \frac{1}{(\phi'(z))^{\frac{3}{2}}}\left(\frac{f(z)}{\sqrt{\phi'(z)}} - \frac{d^2}{dz^2}\frac{1}{\sqrt{\phi'(z)}}\right),$$

where $\phi(z)$ is an arbitrary function which in turn satisfies the fixed point conditions at (4.3). As Appell noted, Koenigs proved that $B'(0) = 1$, thus the fundamental set of solutions are analytic in a neighborhood of the fixed point. However, due to their strict dependence on Koenigs' function $B(z)$, Appell's solutions to the Hill equation are locally defined.

Appell's high regard for the work of Koenigs is evident throughout [1891]. Not only did he explicitly mention the dependence of his approach on Koenigs' "important theorems concerning the existence and general expression of holomorphic solutions to certain functional equations [1891:282]," but in several ancillary remarks he applied other ideas from Koenigs' work to the study of the Hill equation.

4.3 Grévy and the Superattracting Case

Grévy was born in 1865 and died in 1930. The four papers he wrote on functional equations, published between 1892 and 1897, comprise virtually his entire research output. He studied at the École Normale Supérieure and, under the supervision of Koenigs, presented his paper [1894] as his doctoral thesis. From 1897 until the time of his death he taught at the Lyceum Saint Louis in Paris. His thesis [1894] and his paper [1897] form the core of his research on functional equations, and his treatment of the superattracting case, that is, the case where $\phi'(0) = 0$, is largely confined to [1894].

The principal subject of [1894] was not the iteration of complex functions but rather the solution of functional equations, in particular the following one:

$$p_0(z)f(z) + p_1(z)f(\phi(z)) + \cdots + p_n(z)f(\phi^n(z)) = 0, \tag{4.4}$$

where the complex analytic functions $p_i(z)$ and $\phi(z)$ are given, and where $\phi(z)$ has a fixed point at 0 satisfying $0 \leq |\phi'(0)| < 1$. Since it may provide insight into its genesis, it is worthwhile to point out that this functional equation is a generalization of the Schröder functional equation

$$B(\phi(z)) = hB(z), \tag{4.5}$$

which Koenigs solved on a neighborhood of 0 assuming that $h = (\phi'(0))^k$, where k is an integer, and $|\phi'(0)| \neq 0, 1$. Setting $n = 1$, $p_0(z) \equiv h$, and $p_1(z) \equiv -1$, equation (4.4) reduces to the general Schröder equation (4.5).[1]

Grévy used the functional equation (4.4) to treat the $\phi'(0) = 0$ case, and in addition applied it to the solution of several other functional equations. He also observed interesting similarities between the solution of (4.4) and the solution of certain differential equations, a link which Bourlet explored further in his paper on functional operators, [1899a].

The following theorem served as the foundation for Grévy's study of the functional equation (4.4).

Theorem 4.1 (Grévy) *Let $\phi(z)$ be a complex function which is analytic on a neighborhood of 0 and which satisfies both $\phi(0) = 0$ and $0 \leq |\phi'(0)| < 1$. If the function $p_0(z)$ is non-zero at $z = 0$, if*

$$p_0(0) + p_1(0) + \cdots + p_n(0) = 0,$$

and if, in addition, there exists no positive integer m such that

$$p_0(0) + p_1(0)(\phi'(0))^m + p_2(\phi'(0))^{2m} + \cdots + p_n(0)(\phi'(0))^{nm} = 0,$$

then the functional equation (4.4) has an analytic solution $f(z)$ in the neighborhood of 0 which is non-zero at $z = 0$.

That the hypotheses of this theorem allowed $\phi'(0)$ to be 0 gave Grévy the wedge he needed to treat the $\phi'(0)$ case.

Grévy's treatment of the functional equation (4.4) is on the whole rigorous, and the heuristic approach seen in Koenigs' paper [1883], as well as in the work of Schröder and Korkine is absent from Grévy's thesis. One vaguely troubling aspect of his work is its reliance on multi-valued solutions to functional equations. For example, using the above theorem, Grévy showed that the function

$$C(z) = \beta \log z + \sum_{i=0}^{\infty} \beta_i z^i \tag{4.6}$$

[1]Functional equations which have the Schröder equation $f(\phi(z)) = hf(z)$ as a special case were common in the nineteenth century study of iteration. Grévy considered several generalized Schröder equations, and Leau considered a series of simultaneous functional equations, a special case of which is the vector equation $\vec{u}(\phi(z)) = A\vec{u}(z)$, where A is a constant n by n matrix and $\vec{u}(z)$ an n-dimensional complex vector [1897:3].

satisfies the Schröder equation

$$C(\phi(z)) = kC(z) \tag{4.7}$$

where $\phi'(x) = 0$, and k is the order of the first non-zero term in the Taylor expansion of $\phi(z)$ about 0, that is,

$$\phi(z) = \sum_{i=k}^{\infty} \alpha_i z^i$$

with $k > 1$. He defined the function $C(z)$ via a path integral on a deleted neighborhood of 0 of a function with a pole of order 1 at the origin.

As was the case with Koenigs' solution to the Abel equation in the event that $0 < |\phi'(0)| < 1$ (see equation (3.11)), Grévy's solution is multi-valued and satisfies equation (4.7) only in the sense that there is a deleted neighborhood D of the origin such that for each point z_0 in D, a determination of $C(z)$ exists such that equation (4.7) is satisfied at z_0. Just as Koenigs correctly observed that analytic or meromorphic solutions to the Abel equation do not in general exist, Grévy likewise proved that generally no analytic or meromorphic solutions of (4.7) exist.

In the latter portions of his thesis, Grévy applied the function $C(z)$ to the differential equations in much the same manner that Appell applied Koenigs' function $B(z)$ to the study of the Hill equation.

Grévy's function $C(z)$ also can be used to define an analytic solution to the following functional equation which Grévy did not examine, the so-called Böttcher equation

$$f(\phi(z)) = (f(z))^k, \tag{4.8}$$

where, as in equation (4.7), k is the order of the first non-zero derivative of $\phi(z)$ at $z = 0$. The solution $f(z)$ to the Böttcher equation is obtained from equation (4.7) by defining

$$f(z) = e^{C(z)},$$

in which case

$$f(\phi(z)) = e^{C(\phi(z))} = e^{kC(z)} = (f(z))^k.$$

This equation was discussed by Fatou and Julia, who attributed its solution to the mathematician Lucyan Böttcher in the paper [1904], written in Russian. Böttcher was born in Warsaw in 1872. He attended the University of Warsaw but his studies there ended abruptly in 1894 when his participation in a centenary march commemorating the 1794 Polish rebellion against Russia resulted in his expulsion from the university. He then enrolled in a polytechnical institute to study the construction of machines. He evidently kept up his interest in mathematics, however, for he soon moved to Leipzig and received his doctorate from the University of Leipzig in 1898. He subsequently published several works in both Polish and Russian journals.

Contemporary studies of complex dynamics treat the $\phi'(0) = 0$ case via the conjugacy expressed in the Böttcher equation since it presents a better description of iteration near a superattracting fixed point than does Grévy's equation

$$C(\phi(z)) = kC(z).$$

To see this observe that

$$C(\phi^n(z)) = k^n C(z),$$

which suggests that iteration near the attracting fixed point 0 of $\phi(z)$ acts like iteration of the map kz near ∞ since $|C(z)|$ grows without bound as iterates of $\phi(z)$ approach the fixed point 0.

This is not a bad model of iteration: k is the order of the first non-zero higher order derivative of $\phi(z)$, hence it is greater than 1, and points near infinity get closer to infinity under iteration by the linear map $z \mapsto kz$, $k > 1$. However, Grévy's equation is not the best possible model for iteration near a superattracting fixed point. As Schröder pointed out in his paper [1870] (see the end of Section 1.3), if $\phi'(0) = 0$ then iterates of $\phi(z)$ converge to the fixed point 0 not linearly, but rather on the order of z^k, which is precisely the information conveyed by the Böttcher equation

$$f(\phi(z)) = (f(z))^k,$$

which asserts that $\phi(z)$ is conjugate to z^k.

It is curious that Grévy's investigations did not lead him to discover the Böttcher equation. Not only is it, for the reasons outlined above, the natural functional equation to consider in this case, but its solution follows readily from a functional equation Grévy considered. Moreover, given the many parallels between Koenigs' approach and his own, it seems reasonable that Koenigs' realization that the Abel equation could be solved by taking the log of the solution $B(z)$ to the Schröder equation might have suggested to Grévy that the reciprocal technique of exponentiating solutions of known functional equations might yield solutions to other interesting functional equations.

It is important to keep in mind the fact that the aim of Grévy's thesis [1894] was not the definitive treatment of the superattracting case but the definitive treatment of the functional equation (4.4). For this reason it is possible that Grévy did not necessarily intend the functional equation (4.7) as an expression of a conjugacy which accurately modeled iteration. It is more probable that he was following a lead set by Koenigs—perhaps to an extreme—and treated the $\phi'(0) = 0$ case by first solving a Schröder equation and using it in turn to solve other functional equations. As was noted at the end of the previous chapter, Koenigs viewed the extension of his solution of the canonical Schröder equation to the treatment of other functional equations as one of the primary applications of the theory he had developed. That Grévy solved several functional equations regarding the $\phi'(0) = 0$ case, applied them to the study of functional equations and even used Koenigs' solution of the canonical Schröder equation along the way, no doubt led him to believe that he had treated this particular case comprehensively.

Grévy's emphasis on functional equations underlines a dichotomy common in the nineteenth century study of iteration: some mathematicians gravitated towards an interest in functional equations as the primary object of interest, while others evidenced a stronger interest in iteration itself. Grévy certainly falls into the former category. As will be seen in the next chapter, Lémeray falls into the latter category while Leau, like Koenigs before him, balanced the two approaches.

Chapter 5

Iteration in the 1890's: Leau

5.1 Basic Results in the $|\phi'(0)| = 1$ Case

The most troublesome behavior involving fixed points occurs when the derivative at the fixed point, often called the multiplier of the fixed point, has modulus one. Consequently, Koenigs made no headway with this case, and it was not until the mid-1890's that any progress was made.

As the reader will recall, an iterate of a point sufficiently close to a fixed point either eventually moves closer to (i.e., is attracted to) or further away from (i.e., is repelled from) the fixed point according to whether the modulus of the derivative is, respectively, strictly less than or strictly greater than one. This behavior is summarized in figure 5.1.

When the modulus of the multiplier is equal to one the situation is not so clear cut, and a number of different things may occur. I will assume as before that the fixed point is 0. If the multiplier $\phi'(0)$ is a root of unity, then on any neighborhood of the origin $\phi(z)$ exhibits both attracting and repelling behavior, that is, there are some points which under iteration move closer to the origin, and there are some which move away. A more precise description of this behavior appears below in Theorem 5.1 and its corollary.

If, on the other hand, $\phi'(0)$ is one in modulus but not a root of unity then, provided a certain number theoretic condition is satisfied (see Theorem 5.4), iteration acts just like an irrational rotation of the disc in the sense that iterates of a given point sufficiently near the fixed point lie on a topological circle surrounding, but not including, the fixed point. In this event $\phi(z)$ is conjugate to the map $z \mapsto \phi'(0)z$ in a sufficiently small neighborhood of the fixed point. When this particular number theoretic condition is not met, iteration does not necessarily act like an irrational rotation, and in fact, it is generally very difficult to predict the behavior of iterates

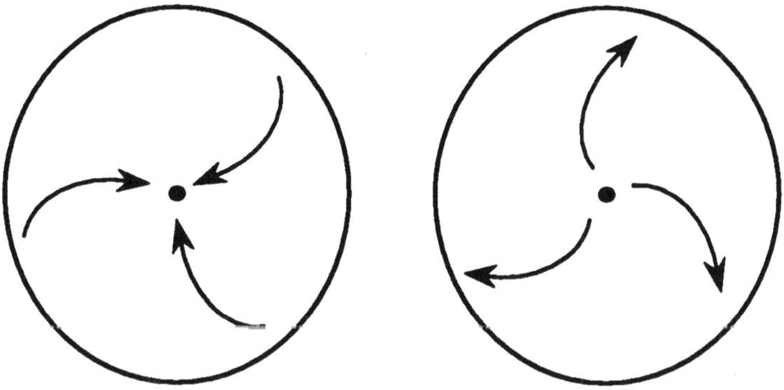

Figure 5.1: *On the left is a representation of an attracting fixed point and on the right, a repelling fixed point. The arrows indicate the direction of movement of iterates near the fixed point. Thus, in a sufficiently small neighborhood of a fixed point x, iteration moves points closer to x if it is attracting and further away if x is repelling.*

near such a fixed point. As of this writing, open questions remain in the case where the multiplier is one in modulus but not a root of unity.

The investigations of both Fatou and Julia regarding the case where the multiplier equals a root of unity were both foreshadowed and influenced by the studies of Leau and Lémeray, although Lémeray's impact on Fatou and Julia was evidently indirect. In what follows, I will first give contemporary renderings of some standard results. I will then review the contributions of both Lémeray and Leau and will conclude my discussion with an analysis of the respective approaches of Fatou and Julia to the $\phi'(0) = 1$ case. A discussion of their work regarding the case where the multiplier is one in modulus but not a root of unity will take place in Chapter 11.

Although their approaches differed, both Fatou and Julia proved equivalent theorems regarding the root of unity case. Their results are summarized below in what is nowadays generally referred to as the Flower Theorem.[1]

Theorem 5.1 (Flower Theorem) *Let the Taylor expansion for $\phi(z)$ around 0 be*

$$\phi(z) = z + \alpha_{m+1}z^{m+1} + \cdots.$$

There exist m disjoint attracting regions, $A_1, \ldots, A_m$, and m disjoint repelling regions, $R_1, \ldots, R_m$.[2] *These regions alternate so that A_i intersects only with R_i and*

[1] Although both Julia and Fatou proved or recognized the individual claims of the Flower Theorem, neither gathered them into a single theorem but rather stated them in a sequence of separate propositions.

[2] A region A is attracting if it is open and connected, and if for all $z \in A$, $\lim_{n\to\infty} \phi^n(z) = 0$.

R_{i+1}. The union of all of these regions and the origin forms a neighborhood of the origin. Each of these regions A_i satisfies $\phi[A_i] \subset A_i$. Moreover, each of the A_i is bisected by a ray emanating from the origin through an mth root of $-1/\alpha_{m+1}$. Analogous statements hold for the repelling regions.

The directions given by the rays in the above theorem are called *attracting directions*. Iterates of all z in a given A_i approach the origin asymptotic to the attracting directions contained in that particular A_i, in the sense that if θ_i is the argument of the attracting direction in A_i, then $\arg(\phi^n(z))$ approaches θ_i as n approaches ∞. The iteration schema for various cases of the Flower Theorem is given in figures 5.2 and 5.3.

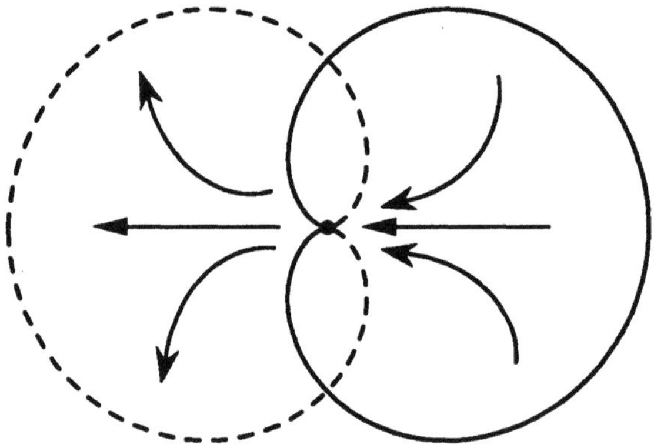

Figure 5.2: *The Flower Theorem in the $m = 1$ case. The arrows indicate the direction of the iterates of $\phi(z)$; iteration under $\psi(z)$ proceeds in a direction contrary to that indicated by the arrows. The repelling petal is outlined with dotted lines.*

The regions A_i or R_i are sometimes called Fatou petals. The use of the terms repelling and attracting, although standard, may be cause for some confusion. According to the Flower Theorem, the intersection of adjacent petals R_i and A_i is not empty, hence a point may be in both a repelling region and attracting region, which suggests that such a point both moves away from the origin and eventually converges to it at the same time. Such a thing can, in fact, happen since iterates of a point z_0 in the intersection of R_i and A_i generally leave R_i before converging to the origin along a path which is asymptotic to the attracting direction in A_i.

The following corollary to the Flower Theorem describes iteration in the case where the derivative at a fixed point is an *nth* root of unity:

An open and connected region R is repelling if, for all $z \in R$, $\psi^n(z)$ converges to 0, where $\psi(z)$ is a local inverse of $\phi(z)$ satisfying $\psi(0) = 0$.

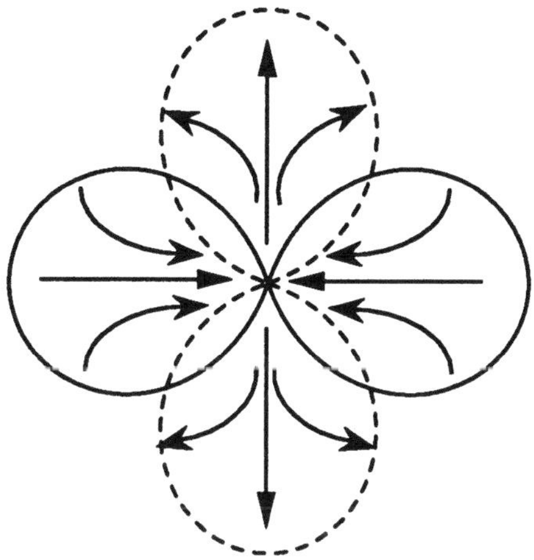

Figure 5.3: *A representation of the Flower Theorem in the $m = 2$ case.*

Corollary 5.2 *When $\phi(z)$ is of the form*

$$\phi(z) = \lambda z + \alpha_{m+1} z^{m+1} + \cdots,$$

where $\lambda^n = 1$, $\phi(z)$ is conjugate to a function of the form $g(z) = \lambda z + z^{nk+1} + \cdots$. Thus $\phi^n(z)$ is conjugate to $g^n(z) = z + \beta_{nk+1} z^{nk+1}$. There are nk attracting Fatou petals and nk repelling Fatou petals for $\phi(z)$. Finally, each of the attracting petals is fixed by $\phi^n(z)$, and the set of nk attracting petals are permuted by $\phi(z)$. If this permutation is denoted by τ, then τ is composed of k cycles of length n.

The thrust of the corollary is that the petal structure for $\phi(z)$ is identical to that of $g^n(z)$. However, the iteration of $\phi(z)$ is slightly different from that of $g^n(z)$, since $\phi[A_i] \subset A_i$ no longer holds. Instead, if $z \in A_i$ then $\phi(z) \in A_{i+jk}$, for some integer j. Thus, iterating by $\phi(z)$ is, roughly speaking, akin to first iterating by $g^n(z)$ and then following it with a rigid rotation.

As noted above, if $|\phi'(0)| = 1$ but $\phi'(0)$ is not a root of unity, then the situation is a bit more complicated. In some cases $\phi(z)$ acts like an irrational rotation of the disk and in others it does not. The criterion for determining whether or not $\phi(z)$ acts like a rotation is number theoretic, and was first established by Carl Siegel (1896–1981) in his paper [1942]:

Definition 5.3 *The quantity $\theta/2\pi$ is not well-approximated by rationals if there exist positive constants a, b such that for all rational numbers p/q*

$$\left| \frac{\theta}{2\pi} - \frac{p}{q} \right| > \frac{a}{q^b}.$$

Theorem 5.4 (Siegel) *Let $\phi(0) = 0$ and $\phi'(0) = e^{i\theta}$. If $\theta/2\pi$ is not well-approximat[ed] by rationals then there exists a locally defined analytic function $f(z)$ solving the Schröder equation*

$$f(\phi(z)) = \phi'(0)f(z).$$

The set of such $\theta/2\pi$ is of full measure on the unit circle.

This theorem implies that if the quantity $\theta/2\pi$ is not well-approximated by rationals then $\phi(z)$ does indeed act like an irrational rotation of the disk. It should also be noted that in the event that $\phi'(0)$ is a root of unity, no solutions to the above Schröder equation exist. For a more detailed discussion of the standard results consult Blanchard [1984], Milnor [1990] or Beardon [1991].

5.2 Lémeray

Leau's thesis [1897] contains the definitive nineteenth century treatment concerning the case where $\phi'(0)$ is an nth root of unity. Indeed, it is the only such study cited by Fatou and Julia. Lémeray, however, treated the $\phi'(0) = 1$ case two years earlier in his paper [1895] and first announced results concerning the root of unity case in his note [1896b]. There is little doubt that he influenced the subsequent study of iteration in the $\phi'(0) = 1$ case, since his work was reviewed and announced in various French mathematical publications in the mid-1890's, and cited in Leau's thesis [1897] as well.

Lémeray, like almost all of those who wrote about the iteration of complex functions in the 1890's, was French. Early in his life he worked as a maritime engineer. He also served as a calculator at an observatory in Algeria and later taught at the College of Dieppe, where he was appointed professor at the age of 32. Besides investigating the case where the derivative of the fixed point is one in modulus, Lémeray published several other short papers on various aspects of iteration including analytic iteration [1898d] and numerical equation solving methods [1898b]. His work also contains an early instance of a technique called graphical analysis which is often used to treat iteration of real functions of one variable (see figure 5.4). In figure 5.4 is a diagram from [1897c] in which he used this technique to illustrate the iteration of a real function $f(x)$ near a fixed point a satisfying $f'(a) = 1$.

Lémeray began his paper [1895] by remarking that in studying the $\phi'(0) = 1$ case he was treating what was, after Grévy's thesis [1894], the last of the cases Koenigs excluded. Lémeray's paper consisted entirely of a demonstration of the following theorem:

Theorem 5.5 (Lémeray) *Let $\phi(z)$ be a complex function of the form*

$$\phi(z) = z + \alpha_{m+1}z^{m+1} + \cdots.$$

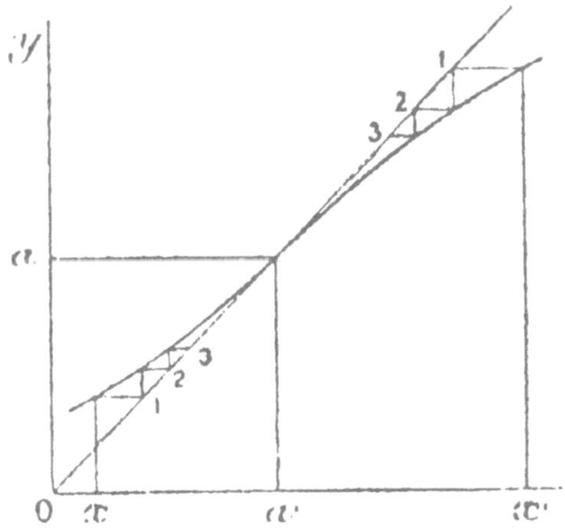

Figure 5.4: *An example of graphical analysis from Lémeray's paper [1897c]. Here, $f(x)$ is real function, $f(a) = a$, $f'(a) = 1$ and $f''(a) = 0$. To follow iterates of x (or x'), follow the vertical line emanating from x (or x').*

If there exists a region A in which all points converge to the fixed point 0 under iteration, then

$$\lim_{n\to\infty} n(\phi^n(z))^m = \frac{-m}{\alpha_{m+1}} \tag{5.1}$$

for all points z in A.

Most of Lémeray's work was carefully argued. However, he occasionally failed to rigorously ground his arguments. For example, the existence of the limit (5.1) is predicated on the existence of an attracting region A, that is, one in which all points converge to 0 under iteration by $\phi(z)$. However, in his paper [1895] he did not bother to prove that such a region actually exists, a shortcoming he did not fully remedy in subsequent papers.

Although Lémeray did not make explicit note of the fact, the limit at (5.1) implies that, if A exists, then for large n

$$\phi^n(z) \approx \sqrt[m]{\frac{-m}{n\alpha_{m+1}}}, \tag{5.2}$$

which in turn suggests that iterates of points in A approach the fixed point asymptotic to the fixed directions given by the mth roots of $-1/\alpha_{m+1}$, which are the attracting directions given by the Flower Theorem (Theorem 5.1). For example, for the function $\phi(z) = z + z^2$, the fact that $m = 1$ and $\alpha_2 = 1$ (5.2) implies that

$$\phi^n(z) \approx \frac{-1}{n}.$$

Consequently, iterates of $\phi(z)$ approach 0 asymptotically to the negative real axis.

Although Lémeray made no explicit mention that (5.1) implies that orbits approach the fixed point asymptotically to the directions given in (5.2), he hinted at the implication by remarking that (5.1) was independent of the choice of z in an attracting region A. Moreover, given the geometric bent of his later papers, it is doubtful that this implication was lost upon him.

Although he still exhibited some carelessness in establishing the existence of regions on which iterates of $\phi(z)$ converge to 0, Lémeray showed in [1897c] that in the case where

$$\phi(z) = z + \alpha_{m+1} z^{m+1} \cdots,$$

a sufficiently small neighborhood D of the origin contains m distinct regions on which $\phi^n(z)$ converges to the fixed point. These regions alternate with m regions on which a local inverse of $\phi(z)$ fixing 0 also converges under iteration to 0. The diagram he used to represent D is given in figure 5.5. Unlike the decomposition given by the Flower Theorem, Lémeray depicted adjacent regions as disjoint, so it is unclear whether he realized in [1897c] that most points in the repelling regions also converge to 0 under iteration by $\phi(z)$, a fact which Leau explicitly recognized. Nonetheless, Lémeray's papers collectively anticipate the Flower Theorem.

In his articles [1896b], [1897e] and [1898a], Lémeray broadened his investigation to include the case where the derivative at the fixed point is a complex root of unity and anticipated the Corollary to the Flower Theorem. He announced in [1897e] a result he proved in [1898a], namely, that if

$$\phi(z) = e^{\frac{p}{n}2\pi i} z + \alpha_{q+1} z^{q+1} + \cdots$$

then

$$\phi^n(z) = z + \beta_{nk+1} z^{nk+1} + \cdots$$

for some integer k. This result enabled him to use the iterative properties of $\phi^n(z)$ to deduce those of $\phi(z)$.

Contrary to the prevailing tradition, Lémeray considered no functional equations in his study of the case where $\phi'(0)$ is a root of unity, although he did study functional equations in a somewhat different context in [1899a]. In further contrast to the majority of those who studied iteration in the 1800's, Lémeray often sprinkled his papers with diagrams, two of which I have included.

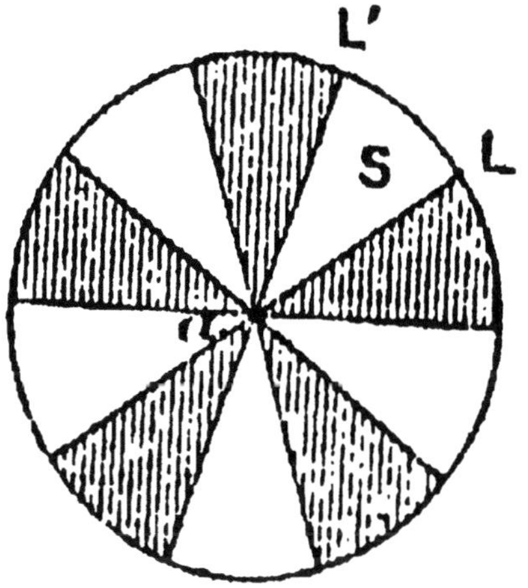

Figure 5.5: *A diagram from Lémeray's paper [1897c] depicting the division of a disc surrounding a fixed point a into disjoint regions of convergence and divergence where $\phi'(a) = 1$ [1897c:313]. According to Lémeray, iterates converge to the fixed point a on the non-shaded regions.*

5.3 Leau's Work

Leau studied at the École Normale Supérieure and received his doctorate in April 1897. He spent most of his professional life teaching at the University of Nancy and served there as Dean of the Faculty of Science from 1931-34. He published a modest number of research papers over the course of his life, none of which were as influential as his thesis.

Leau's thesis [1897] was the most important work on iteration of complex functions to appear in the 1890's. In answering many of the remaining questions about iteration near a fixed point, Leau's thesis is also the last major French work to concern itself with the local study of iteration. Although Lémeray and Grévy also explored aspects of iteration that Koenigs did not, neither influenced further study to the extent that Leau did. Leau's thesis is also impressive for its breadth. Not only did he study iteration in the case where $\phi'(0)$ is a root of unity, but he also investigated analytic iteration, iteration of functions of more than one variable, as well as systems of simultaneous functional equations. Leau's thesis, unfortunately, offers a few problems along with its delights. His reasoning is occasionally obscure, and his descriptions of results are sometimes confusing. These defects are present in his treatment of what is the most important theorem of his thesis, Theorem 5.6, a nascent version of the Flower Theorem.

Although much of Lémeray's work predated Leau's thesis, Leau's influence quickly eclipsed that of Lémeray, whose contributions to the study of complex dynamics have been largely forgotten. Neither Fatou nor Julia mention Lémeray, and both incorrectly credit Leau with having initiated the study of the $\phi'(0) = 1$ case. Perhaps the reason that Lémeray has faded into obscurity is that Leau's work not only duplicated and extended Lémeray's studies, but is also closer to the prevailing trends of nineteenth century iteration theory since it included treatments of various functional equations. Moreover, in contrast to Lémeray, who worked in Dieppe, Leau worked amidst the French mathematical mainstream in Paris, and the fact that his thesis was written under the guidance of Koenigs certainly did little to hinder its dissemination.[3]

5.4 Leau's Anticipation of the Flower Theorem

Like Lémeray before him, Leau treated the case where

$$\phi(z) = e^{\frac{p}{n}2\pi i}z + \alpha_{m+1}z^{m+1} + \cdots$$

by showing that

$$\phi^n(z) = z + \alpha_{nk+1}z^{nk+1} + \cdots.$$

As noted in the discussion following Corollary 5.2, this effectively reduces the case of a complex root of unity to that of $\phi'(0) = 1$, followed by a rotation. Consequently, in the remainder of this chapter, only the $\phi'(0) = 1$ case will be considered.

Leau's anticipation of the Flower Theorem (Theorem 5.1) is equivalent to the following:

Theorem 5.6 (Leau) *Let $\phi(z)$ have a Taylor expansion of the form*

$$\phi(z) = z + \alpha_{m+1}z^{m+1} + \cdots.$$

Let $\psi(z)$ be a local inverse of $\phi(z)$ satisfying $\psi(0) = 0$. There exists a disc D surrounding 0, such that all $z \in D$ converge to 0 under iteration by $\phi(z)$, $\psi(z)$ or both functions.

The statement of this theorem has little of the geometric richness of the Flower Theorem. In his proof, however, Leau claimed that in the $m = 1$ case, that is, where

$$\phi(z) = z + \alpha_2 z^2 + \cdots,$$

[3] Appell was the chairman of his examination committee while Émile Picard and Koenigs were the other members. Picard and Appell, it is worth noting, served on the commission of the French Academy of Sciences that awarded the 1918 *Grand Prix de Mathématiques* to Julia. Koenigs' influence on Leau is underscored by the fact that Leau dedicated his thesis to him.

A and R are both cardioids, which is precisely the decomposition given by the Flower Theorem (see figure 5.2). He included a sketch of such a cardioid in [1897], which is reproduced in figure 5.6. Leau said little about the structure of these regions in the general case except to remark that they may consist of disjoint pieces "and may have a greater extent than one might realize [1897:33]."

Nonetheless, in his justification of Theorem 5.6, Leau treated a special family of functions, $\mathcal{F}$, given below at equation (5.3), whose properties under iteration might well have suggested to Fatou and Julia the decomposition summarized in the Flower Theorem. It was Leau's intention to use functions from the family $\mathcal{F}$ to approximate iteration by arbitrary functions. Although he failed to state his case with sufficient detail and clarity, a fully detailed proof could be based on his approach. Indeed, Julia's proof of the Flower Theorem is in many respects a modification of Leau's approach (see Section 6.1).

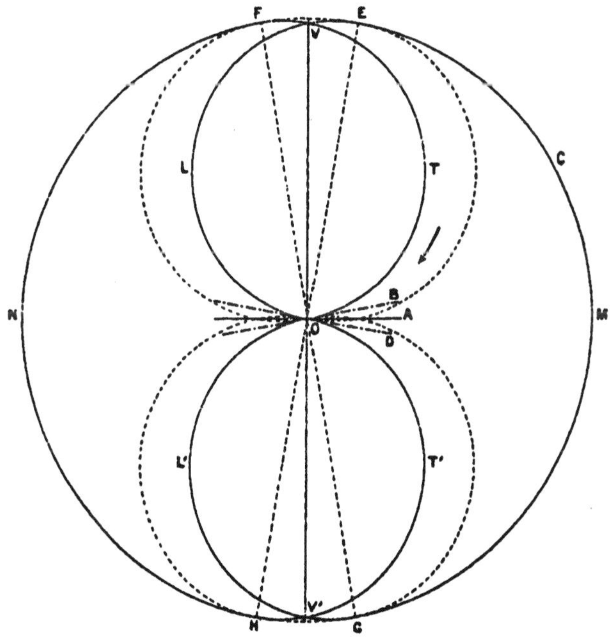

Figure 5.6: *Above is a diagram from [1897:32]. Leau claimed that the function $\phi(z)$ converges to 0 under iteration interior to the cardioid $OLML'$. Iterates of $\phi^{-1}(z)$ converge to 0 interior to the cardioid $OTNT'$. The arrow indicates the direction in which forward iterates of $\phi(z)$ travel for points in the upper-half of the interior of the cardioid $LOML'$.*

The family of functions $\mathcal{F}$ referred to above consists of the functions $\lambda(z)$ determined by

$$(\lambda(z))^m = \frac{z^m}{1 - hz^m}, \tag{5.3}$$

where m is a positive integer and h a non-zero complex constant. Leau resolved the ambiguity involved in determining $\lambda(z)$ by requiring that $\lambda'(0) = 1$. This constraint is quite reasonable since a routine calculation shows that the derivative at 0 of any of the mth roots of $(\lambda(z))^m$ must be a root of unity.

Given a fixed function of the form

$$\phi(z) = z + \alpha_{m+1} z^{m+1} + \cdots,$$

Leau in essence set $h = m\alpha_{m+1}$, where the m in $(\lambda(z))^m$ is the same m used in the above expression for $\phi(z)$. The function $\lambda(z)$ is then analytic as long as $z^m \neq 1/h$, hence there exists a neighborhood D of 0 on which $\lambda(z)$ is analytic. Leau then approximated iteration of $\phi(z)$ in D by iterating $\lambda(z)$ and analytically perturbing h slightly as the iterates $\phi^n(z)$ vary over D in order to refine his approximation.

Although Leau did not fully explain his reasons for using this approximation, three reasons suggest themselves. First of all, both $\phi(z)$ and $\lambda(z)$ are of the form

$$z + Az^{m+1} + \cdots,$$

where A is a non-zero constant. Moreover, since $h = m\alpha_{m+1}$, $\lambda(z)$ and $\phi(z)$ have the same attracting directions, that is, iterates of both functions approach the origin asymptotic to the directions

$$\sqrt[m]{-\frac{1}{\alpha_{m+1}}},$$

which are given by the Flower Theorem and implied by Lémeray's Theorem 5.5.[4] Finally, the iteration schema for the functions in $\mathcal{F}$ is relatively simple. As will be seen below, if $m = 1$ and h is held constant, then $\lambda(z)$ fixes a certain family of circles; that is, for a circle γ in this family of circles, $\lambda[\gamma] \subset \gamma$. Therefore, for a point z on γ, iterates of $\lambda(z)$ remain on γ. In the general case, again for fixed h, $\lambda(z)$ leaves invariant a certain family of $2m$-leaf roses (see figure 5.7).

To create the cardioid $OLML'$ pictured at figure 5.6 for the function

$$\phi(z) = z + \alpha_2 z^2 + \cdots,$$

Leau, in essence, set

$$\lambda(z) = \frac{z}{1 - \alpha_2 z}, \tag{5.4}$$

[4]Leau was aware of Lémeray's theorem as well as the existence of attracting directions. He not only gave an improved proof of Lémeray's theorem, but explicitly acknowledged that this result implied that iterates must travel towards the origin along "certain lines [1897:34]."

and perturbed α_2 slightly in order to approximate iteration of $\phi(z)$ by iteration of $\lambda(z)$. Leau's discussion of his perturbation is both obscure and complicated. In order to indicate how he constructed the cardioid and at the same time avoid burdensome detail, I will refrain from perturbing α_2 and show how to construct a cardioid interior to which all points converge to 0 under iteration by (5.4).

Using properties of linear fractional transformations, it is not hard to show that the function $\lambda(z)$ fixes a family of circles tangent at the origin to the line through the attracting direction $-1/\alpha_2$, which in figure 5.6 is the line through $M, A, 0$ and N, hereafter denoted MA. Since both $\phi(z)$ and $\lambda(z)$ share the same attracting direction, points on these circles converge under iteration by $\lambda(z)$ to 0 along a path tangent to MA.

The circle OL, which is the circle formed by the union of the solid semi-circle OLE and the dotted semi-circle EO in figure 5.6, is one such circle, as is the circle OL' formed by the solid semi-circle $OL'G$ and the dotted semi-circle GO. Since all circles fixed by $\lambda(z)$ are themselves foliated by such circles, points interior to these circles also converge to 0 under iteration by $\lambda(z)$. Iterates $\lambda^n(z)$ of points z on one of the fixed circles above MA converge to 0 in a clockwise manner; iterates of points below MA do so in a counterclockwise direction.

The cardioid $OLML'$ (and its interior) is then formed by first rotating the circle OL clockwise until it sits atop OL', and then taking the union of all these intermediary circles (as well as their interiors) between OL and OL'. Each point interior to $OLML'$ converges to 0 under iteration by $\lambda(z)$ since it lies on one of the circles tangent to MA which are fixed by $\lambda(z)$. The cardioid $OTNT'$ is formed analogously, and all points interior to it converge to 0 under iteration by $\lambda^{-1}(z)$.

It was Leau's intention to perturb h slightly and use the functions $\lambda(z)$ so created to approximate iteration by $\phi(z)$ and thereby demonstrate that all points interior to the cardioid $OLML'$ also converge under iteration by $\phi(z)$, to 0. Unfortunately, he did not justify his contention with the detail and clarity one might wish.

One of the ways in which Julia's proof of the $m = 1$ case improved upon Leau's investigation was to dispense with the latter's technique of using iterates of the function $\lambda(z)$ to approximate iteration by $\phi(z)$ and offer a direct proof that points interior to an equivalent family of circles converges to 0 under iteration by $\phi(z)$. Moreover, Julia carefully and explicitly generalized his own argument to the $m > 1$ case whereas all Leau did was briefly describe how the family $\mathcal{F}$ behaved and then make the debatable claim that his argument "extends by itself" to arbitrary m [1897:31].

When $m > 1$, Leau observed that for fixed h, the function $\lambda(z)$ leaves invariant a family of $2m$-leaf roses.[5] This family of roses, denoted $\mathcal{R}$, foliates a sufficiently small neighborhood of the origin, as indicated in figure 5.7. Each family of leaves from

[5] This follows from the $m = 1$ case, wherein $\lambda(z)$ fixes a circle, because $z/(1-hz)$ is conjugate to $z^r/(1-hz^r)$ via the map $z \mapsto z^r$ and sets which are invariant under iteration are preserved by conjugation. Although he did not explicitly note this conjugation, it is perhaps because of it that he felt his argument extended "by itself."

Figure 5.7: *The functions $\lambda(z)$ leave a family of $2m$-leaf roses invariant under iteration. The $m = 2$ case is presented here. The arrows indicate the direction in which the iterates travel along the leaves of a particular rose. The family of leaves γ referred to in the text is the family of leaves in the upper-right quadrant, and the family γ_1 is the family in the lower-right quadrant.*

$\mathcal{R}$ is sandwiched between an attracting direction of $\lambda(z)$ and a repelling direction. These roses can also be decomposed into m pairs of adjacent leaves, each of which is separated by an attracting direction. In figure 5.7, the family of leaves in the upper-right quadrant, which will be referred to as γ, and the family in the lower-right quadrant, called γ_1, form such a pair. Moreover, points on both γ and γ_1 converge to 0 under iteration by both $\lambda(z)$ and an appropriately chosen inverse.

While Leau did not do so, the attracting and repelling Fatou petals given by the Flower Theorem can be obtained without too much difficulty from the family $\mathcal{R}$. For example, an attracting Fatou petal A_i for the function $\lambda(z)$ can be formed by first selecting one of the pairs of adjacent families of leaves from the roses $\mathcal{R}$ which are separated by an attracting direction (for example, the leaves γ and γ_1) and then enclosing the pair with a curve Γ as indicated in figure 5.8. Since all points interior to Γ lie on a leaf in $\mathcal{R}$, all points inside Γ converge to 0 under iteration by $\lambda(z)$. Therefore the interior of Γ is an attracting Fatou petal A_i for $\lambda(z)$. A repelling petal is formed in an analogous manner from adjacent families of leaves separated by a repelling direction.

It is not unreasonable to suspect that Fatou and Julia each deduced the structure of the attracting and repelling petals from Leau's discussion in much the same manner, and then went about their respective ways developing a proof. As in the $m = 1$ case, Julia's proof, although along the same lines as Leau's, dispensed with the approximating functions and directly formed the attracting and repelling petals given in the Flower Theorem. Fatou's proof, on the other hand, does not appear

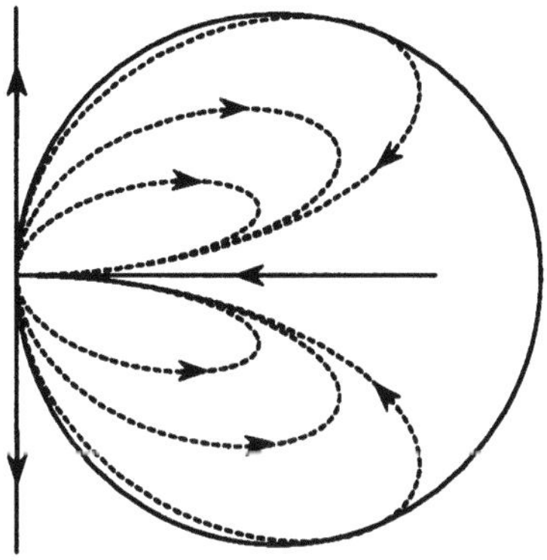

Figure 5.8: *The families referred to in the text as γ_1 and γ can be enclosed in a curve Γ whose interior is an attracting Fatou petal for $\lambda(z)$.*

to be rooted in Leau's approach, yet he nevertheless depicted the attracting and repelling regions in the $m = 1$ case as cardioids and used a petal structure identical to that given in the Flower Theorem for the general case. Before discussing Fatou's and Julia's proofs, I will briefly examine Leau's use of functional equations.

5.5 Leau and Functional Equations

Although Leau at times seemed to solve functional equations in a *pro forma* bow to tradition, he also used them in an attempt to obtain a geometric picture of iteration in the neighborhood of a fixed point 0 in the general case where $|\phi'(0)| \leq 1$. He did this by constructing arcs $\gamma(t)$ which are invariant under iteration by $\phi(z)$, that is, $\phi[\gamma(t)] \subset \gamma(t)$. These arcs were implicitly defined via an analytic iteration function $\Phi(t, z)$ of the sort used by Schröder (see equation (2.5) in Section 2.2), where t is a non-negative real number and z is complex.

Leau constructed these arcs $\gamma(z)$ in all cases except the case in which the derivative of the fixed point was one in modulus but not a root of unity. He did so via functional equations solved by Grévy in the $\phi'(0) = 0$ case (equation (4.7)), Koenigs in the $0 < |\phi'(0)| < 1$ case (the canonical Schröder equation) and himself in the case where $\phi'(0) = 1$. In the last case, Leau solved the Abel equation $f(\phi(z)) = f(z)+1$, not on a neighborhood of 0, but in sufficiently small neighborhoods of points near 0, thereby avoiding multi-valued solutions to the Abel equation.

In his construction of these arcs Leau was forced to confront the fundamental shortcoming of the nineteenth century study of iteration, namely, its local nature. In each instance, he defined the arc $\gamma(t)$ in terms of solutions to functional equations which, without exception, were defined locally. As Leau explicitly noted, the curves $\gamma(t)$ could therefore only be constructed locally [1897:67].

Leau's failure to establish a global approach, like Koenigs' failure before him, did not stem from a lack of interest. As was noted during the discussion of his work, Koenigs lamented his inability to divide the plane into regions of convergence, that is, into regions where each point converges under iteration by $\phi(z)$ to the same attractive orbit, and expressed doubts that such a division was even possible. In sentiments which echo the pessimism of Koenigs, Leau noted that

> The problem of extending the solutions [to functional equations] reduces to the problem of the division of the plane into regions which are transformed into one another [which is] without a doubt impossible in the general case ... [1897:55].

Chapter 6

The Flower Theorem of Fatou and Julia

6.1 The Approaches of Fatou and Julia

Julia studied the case where $\phi'(0) = 1$ in his lengthy *Mémoire sur l'iteration des fonctions rationnelles* published in 1918, which was his principal work on the theory of iteration. Fatou discussed this case in the even longer *Sur les èquations fonctionnelles* which was published in three parts in 1919 and 1920. Each of these works represents a fresh and innovative approach to the study of iteration. Although the work of Fatou and Julia will be discussed at length in Chapter 11, I will discuss their contributions to the $\phi'(0) = 1$ case in the present chapter, somewhat out of chronological order. However, before discussing their respective approaches to the $\phi'(0) = 1$ case, it will be worthwhile to say a few words regarding the scope of their studies of iteration.

Unlike the works of their predecessors, Fatou and Julia were able to describe the iteration of arbitrary complex functions beyond the neighborhood of a fixed point. Their success in developing such a global approach was due in large measure to their application of the theory of normal families, which Paul Montel (1876-1975) articulated in a series of papers published in the second decade of the twentieth century (see Chapter 8). Their reliance on Montel's work also accounted for a certain similarity of approach, despite their considerable stylistic differences.

One instance where this similarity in approach did not hold was in their discussion of the $\phi'(0) = 1$ case. This difference is due not only to the fact that the Flower Theorem does not call for the theory of normal families, but also because Fatou and Julia looked at this particular case from substantially different vantage points. Where Julia apparently drew upon Leau's work, Fatou took an entirely fresh approach.

6.2 Julia's Proof of the Flower Theorem

As noted above, Leau's anticipation of the Flower Theorem, Theorem 5.6 above, asserts that if

$$\phi(z) = z + \alpha_{m+1} z^{m+1} + \cdots,$$

then a sufficiently small neighborhood of 0 can be decomposed into sets A and R such that all points on A converge to 0 under iteration by $\phi(z)$, and all points on R converge to 0 under iteration by an appropriately chosen inverse of $\phi(z)$.

As the Flower Theorem asserts, Julia rigorously showed that A and R are cardioids in the $m = 1$ case, and that when $m > 1$, A and R each consists of m petals. His construction of A in the $m = 1$ case is as follows.

Given a function $\phi(z)$ of the form

$$\phi(z) = z + \alpha_2 z^2 + \cdots,$$

Julia constructed the region A by finding sufficient conditions that a point β near, but not equal to, 0 is surrounded by an open disc on which all points converge to 0 under iteration by $\phi(z)$. All points in the union of these discs taken over all such β consequently converge to 0. It will be seen that this union forms a cardioid-like region.

Julia's first step was to define the open disc

$$\Delta_\beta = \{z : |z - \beta| < |\beta|\},$$

and seek conditions on β for which $\phi[\Delta_\beta] \subset \Delta_\beta$. To this end, he proved the following theorem.

Theorem 6.1 (Julia) *Let* $\arg(\alpha_2) = \theta$. *Then there exists a* $\rho > 0$ *such that if* $|\beta| < \rho$ *and* $\arg(\beta) = \omega$ *satisfies*

$$\frac{\pi}{2} - \theta < \omega < \frac{3\pi}{2} - \theta, \tag{6.1}$$

then $\phi[\Delta_\beta] \subset \Delta_\beta$.

Julia next used the following generalization of Schwarz's Lemma, which he also proved in [1918], to show that if $\phi[\Delta_\beta] \subset \Delta_\beta$, then all $z \in \Delta_\beta$ converge to 0 under iteration by $\phi(z)$ [1918:150–52]:

Theorem 6.2 *Let* $\phi(z)$ *be an analytic function such that on a disc* D,

$$\phi[D] \subset D.$$

Suppose as well that in the closure of D, *there is a unique fixed point* x *of* $\phi(z)$ *such that* $0 < |\phi'(x)| \leq 1$. *Then all points in* D *converge to* x *under iteration by* $\phi(z)$.

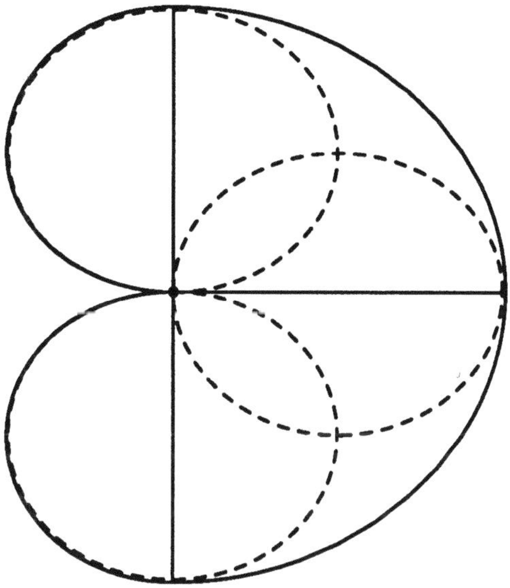

Figure 6.1: *Taking the union of the discs Δ_β, indicated by the dotted curves, Julia formed a cardioid-like region A on which iterates of $\phi(z)$ converge to* 0.

Julia actually used this generalization of Schwarz's Lemma a number of times in [1918]. This generalization is not at all necessary to the study of iteration—for example Fatou did not use it— but it is quite useful. Moreover, Julia's use of it is illustrative of his incorporation of general results from the theory of functions.

Julia observed that the union of the discs Δ_β, where β satisfies the hypothesis of Theorem 6.1, forms a figure which he termed "reminiscent of a cardioid," interior to which all points converge to 0 under iteration by $\phi(z)$ [1918:288]. This construction is indicated in figure 6.1.

It certainly seems quite likely that Julia's approach was influenced by Leau's construction. Not only did Julia cite Leau's work, but both constructed their cardioids by taking a union of a family of discs. Of course, the families involved differed, and Julia's construction was far more direct and precise. Nonetheless, the actual cardioids are virtually identical and their method of construction quite similar.

This similarity persists in the $m > 1$ case. Let $\arg(\alpha_{m+1}) = \theta$ and let ω satisfy

$$\frac{\pi}{2} - \theta < m\omega < \frac{3\pi}{2} - \theta.$$

Julia then showed that in each of the m sectors so defined, a family of open ovals bisecting the ray $\arg(z) = \omega$ can be constructed interior to which all points converge to 0 under iteration by $\phi(z)$. The union of these ovals taken over all ω in a given sector forms one of the m attracting petals called for by the Flower Theorem (see figure 6.2). The repelling regions are constructed analogously by considering the inverse $\psi(z)$ rather than $\phi(z)$.

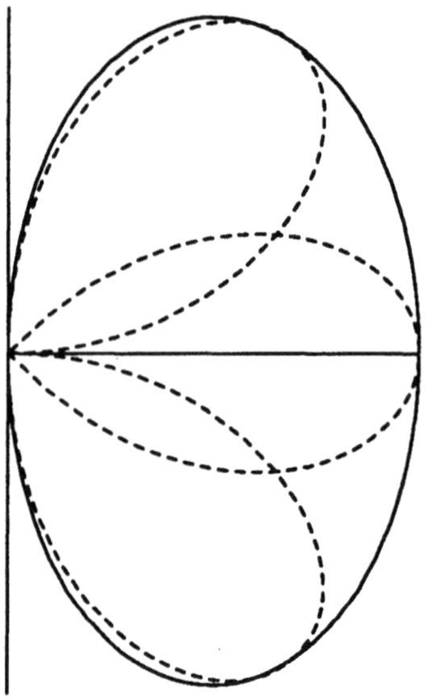

Figure 6.2: *Julia formed an attracting petal A_i in the $m = 2$ case by taking the union of a family of ovals.*

6.3 Fatou's Proof of the Flower Theorem

Fatou constructed the attracting petals A_i for the function

$$\phi(z) = z + \alpha_{m+1} z^{m+1} + \cdots$$

by mapping a certain sector of angle $2\pi/m$ from a neighborhood N of the fixed point 0 onto a neighborhood of N^* of ∞ via the following conjugacy which maps the fixed point 0 to ∞:

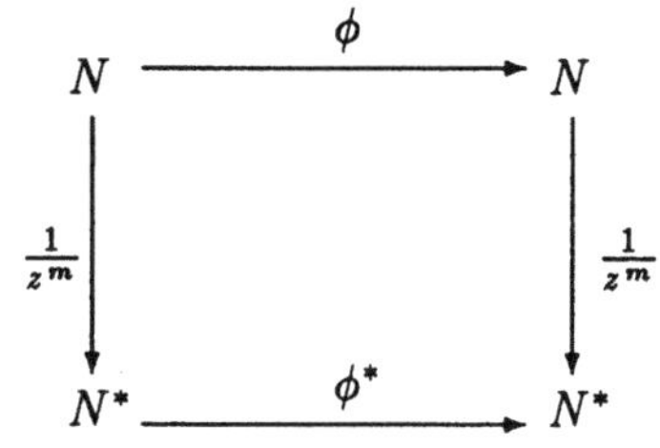

Fatou then showed that iterates of $\phi^*(z)$ converge to ∞ on a subset D^* of N^*, and formed the attracting petal A_i of $\phi(z)$ by mapping D^* back to the origin via an appropriate inverse of $1/z^m$. The technique of treating a fixed point at ∞ by mapping it to 0 via the map $z \mapsto 1/z$ had been suggested by Koenigs in his paper [1883]. Fatou's proof thus relied on a novel twist of this old technique.

In the $m = 1$ case Fatou mapped a neighborhood N of the origin to a neighborhood N^* of ∞ via the map $z \mapsto 1/z$. Hence if $\phi(z)$ is of the form

$$\phi(z) = z + \alpha_2 z^2 + \cdots,$$

then the function $\phi^*(z)$ is of the form

$$\phi^*(z) = \frac{1}{\phi(1/z)}$$

which yields

$$\begin{aligned}\phi^*(z) &= z - \alpha_2 + \frac{\beta_1}{z} + \frac{\beta_2}{z^2} + \cdots \\ &= z + a + g(z),\end{aligned}$$

where $a = -\alpha_2$, and $g(z)$ is the function

$$g(z) = \sum_{i=1}^{\infty} \frac{\beta_i}{z^i}.$$

It may be assumed without loss of generality, by rotating the plane by $\arg(\alpha_2)$ if necessary, that a is a positive real quantity.

Since $\lim_{z\to\infty} g(z) = 0$, the function $g(z)$ is analytic on N^* of ∞, and the maximum modulus u of $g(z)$ occurs on the boundary of N^*. Shrinking N^* if necessary, it can be assumed that $|g(z)| < u < a$ for all $z \in N^*$.

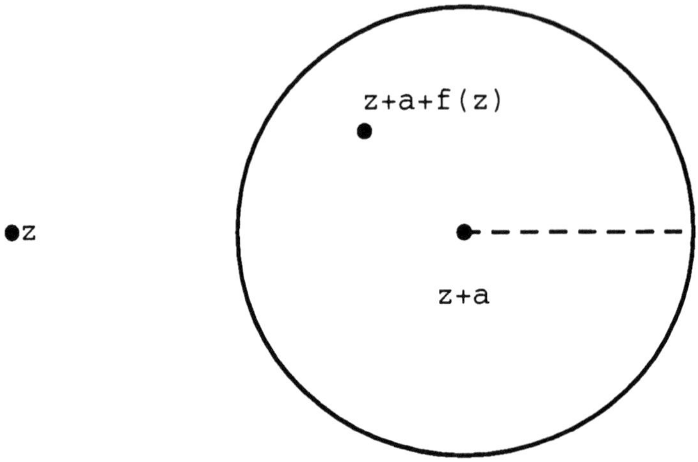

Figure 6.3: *The disc pictured has center $z + a$ and radius u. Since $\phi^*(z) = z + a + g(z)$, and $|g(z)| < u < a$, $\phi^*(z)$ must lie in the interior of the disc. Thus, $\phi^*(z)$ is always to the right of z for all z in N^*.*

The function $\phi^*(z)$ has a fixed point at ∞. Iteration by $\phi^*(z)$ on N^* acts quite a bit like horizontal translation by a. To see this, let $z \in N^*$. Since

$$\phi^*(z) = z + a + g(z),$$

and $g(z)$ approaches 0 as z approaches ∞, the effect of the $g(z)$ term is negligible for large z, in which case $\phi^*(z) \approx z + a$.

If $\phi^*(z)$ is also in N^*, then

$$\begin{aligned}\phi^{*2}(z) &= \phi^*(z) + a + g(\phi^*(z)) \\ &= z + 2a + g(z + a + g(z)),\end{aligned}$$

and the effect of $g(z)$ is even less, since $|z + a + g(z)| > z$ due to the fact that on $N^*(z)$, $a > |g(z)|$. Continuing in this manner we get,

$$\phi^{*n}(z) = z + na + g[z + (n-1)a + \sum_{i=1}^{n} g(\phi^{*i-1}(z))], \tag{6.2}$$

provided $\phi^{*i}(z) \in N^*$ for $i \leq n$.

There is no guarantee however, that $\phi^{*i}(z)$ is in N^*. But by making a geometric observation, a half-plane Δ^* contained in N^* can be found in which $\phi^*(z)$ is in Δ^* whenever z is. Since $|g(z)| < u$ on $N^*(z)$, the fact that $\phi^*(z) = z + a + g(z)$ implies that $\phi^*(z)$ is interior to a disc of radius u centered at $z + a$ (see figure 6.3). Since

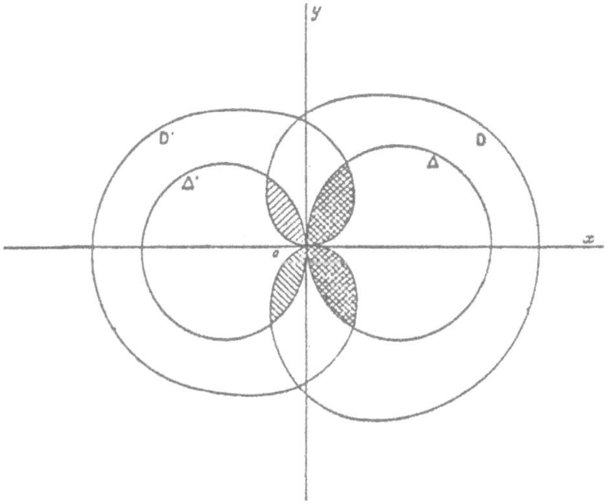

Figure 6.4: *The region interior to the cardioid D is the image of the region D^* exterior to the parabolic curve under $z \mapsto 1/z$ depicted in the previous figure. The disc Δ interior to D is the image of the interior of Δ^* under $z \mapsto 1/z$. The regions D' and Δ' are the corresponding repelling regions. This diagram is from [Fatou 1919:205].*

$u < a$, it also follows that $\phi^*(z)$ is always to the right of z on the extended complex plane. Using these facts, a complex quantity $w \in N^*$ with $\text{Re}(w) > 0$ can be found such that if $\text{Re}(z) > \text{Re}(w) = r$, then $\text{Re}(\phi^*(z)) > r$. Thus if Δ^* is defined as the half-plane where $\text{Re}(z) > r$, it follows that if $z \in \Delta^*$ then $\phi^*(z)$ is as well, hence $\phi^*[\Delta^*] \subset \Delta^*$. Mapping D^* back to the origin via $z \mapsto 1/z$ produces a cardioidal region D with 0 as its cusp point. Fatou's representation of D and Δ are given in figure 6.4.

It can also be shown that the sum in equation (6.2) is negligible. Thus, since na goes to ∞ with n, it follows that $\phi^{*n}(z)$ converges to ∞ for all $z \in \Delta^*$. In other words, for large n, $\phi^{*n}(z)$ is approximated by $z + na$ and iteration by $\phi^*(z)$ amounts to repeated translation by a.

The image under $z \mapsto 1/z$ of the set Δ^* with the point at ∞ removed is an open disc Δ with 0 as boundary point. Since points in Δ^* converge to ∞ under iteration by $\phi^*(z)$, it follows that all points in Δ converge to 0 under iteration by $\phi(z)$. Fatou's construction of Δ is therefore analogous to Julia's construction of a disc Δ_β on which $\phi^n(z)$ converges to 0.

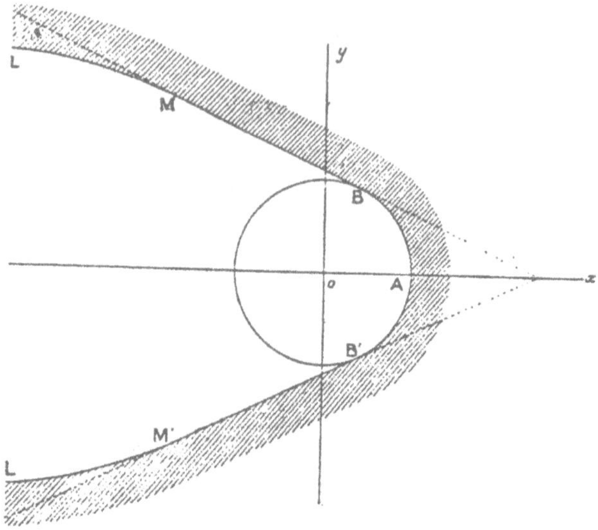

Figure 6.5: *Fatou formed the attracting region D by mapping the region D^* exterior to the parabolic curve $M'B'BM$ into a neighborhood of the origin via the map $z \mapsto 1/z$. This diagram is from [Fatou 1919:194].*

Unlike Julia, however, Fatou did not construct his cardioid by taking a union of discs. Instead, Fatou observed that Δ^* is actually a rather conservative estimate of the set of points which converge to ∞ under iteration by $\phi^*(z)$. He was therefore able to extend this set from Δ^* to a region D^*. The region D^* is depicted in figure 6.5.

Fatou generalized this approach to the case where $m > 1$. The principal difference in this instance is that D^* maps back to an attracting petal A_i given in the statement of the Flower Theorem.

Although Julia avoided functional equations in [1918], apparently as a reaction to what he considered an undue emphasis on them in previous studies, Fatou's investigation of the $\phi'(0) = 1$ case led him to the Abel equation

$$f^*(\phi^*(z)) = f^*(z) + a, \tag{6.3}$$

where a is the constant used above in the expression

$$\phi^*(z) = z + a + g(z). \tag{6.4}$$

It is interesting that Julia avoided functional equations in this particular case since he strove for a certain modernity in his work, and the Abel equation has turned out to be very useful in the contemporary study of fixed points with multiplier one.

In any event, Fatou's treatment of the Abel equation is based on the approximation $\phi^*(z) \approx z + a$ for large z, which, as was noted above, suggests that iteration of $\phi^*(z)$ is very much like repeated translation by a. This in turn suggests the Abel equation, since an analytic solution $f^*(z)$ to equation (6.3) would reduce iteration to repeated translation.

In equation (6.4) the function $g(z)$ approaches 0 as z approaches ∞. This enabled Fatou to construct an analytic function $f^*(z)$ satisfying equation (6.3) on the interior of the set D^* defined above. Mapping this solution back to the attracting region D, he was able to solve

$$f(\phi(z)) = f(z) + a$$

interior to D.

In light of the attention given to the problem of analytic iteration in the nineteenth century study of iteration (see Section 2.2), it is worth noting that Fatou observed that his solution of the Abel equation led to the solution of "the problem of analytic iteration" since on the disc Δ contained in D, the function $f(z)$ is one-to-one, and therefore [1919:203]

$$\phi^w(z) = f^{-1}(f(z) + wa).$$

Chapter 7

Fatou's 1906 Note

7.1 Introduction: Local Versus Global Studies of Iteration

Previous to Pierre Fatou's note "*Sur les solutions uniformes de certaines équations fonctionnelles,*" which appeared in the *Comptes rendus* of the French Academy of Sciences in 1906, studies of iteration focused on a given analytic function $\phi(z)$ in the vicinity of an attracting fixed point.[1] Although much was known about the behavior of $\phi(z)$ under iteration near a fixed point, little was known about the global behavior of such functions, that is, the behavior of the iterates of an arbitrary point in the extended complex plane.

What was known about the global behavior was limited to various special cases. For example, Cayley and Schröder independently classified the iterates of arbitrary points for the Newton's method function $N(z)$ for the complex quadratic $q(z)$. Both men showed that if the roots of $q(z)$ are distinct then all points on a particular side of the perpendicular bisector L of the roots converge under iteration by $N(z)$ to the root on that side of L. They also showed that L is invariant under iteration by $N(z)$, that is, that $N[L] \subset L$.[2] In addition to the studies of Cayley and Schröder, the global iterative properties of a linear fractional transformation (LFT) are easily determined and had been known for quite some time. The global description of the behavior of an LFT under iteration relied upon properties unique to linear maps, such as the fact that the n-fold composition of an LFT with itself is still an LFT, and consequently shed little light on the iteration of other functions.

---

[1] An attracting fixed point x satisfies $|\phi'(x)| < 1$, and there exists an open disc D surrounding x on which $\lim_{n\to\infty} \phi^n(z) = x$ for all z in D. Koenigs studied the behavior of a function $\phi(z)$ in the vicinity of its fixed and periodic points in great detail in his papers [1883], [1884] and [1885]. His work is discussed in Chapter 3.7.

[2] See Chapter 1 for a detailed account of contributions of Cayley and Schröder.

Fatou's note [1906a] represents the first global study involving the iteration of a general class of complex functions, namely, rational functions whose only attracting orbit is a fixed point.[3] This restriction to the consideration of rational functions, that is functions which are the quotient of two polynomials, is in keeping with his emphasis on the global, since rational functions are the only functions which are analytic on the entire extended plane.

Fatou showed that under certain hypotheses, the iterates of a rational function with a unique fixed point converge to that fixed point on the entire extended complex plane except for a totally disconnected perfect set J.[4] Fatou's proof of this fact relied heavily on set theory and signaled the introduction of sophisticated set theoretic techniques into the study of iteration. It should be emphasized, however, that even though this result completely describes the iteration of a wide range of functions, the functions he studied are of a specific nature since they have a unique attracting fixed point.

7.2 The Lack of Set Theory in Koenigs' Work

The lack of knowledge in the late nineteenth century regarding the global properties of iteration did not reflect a lack of interest. As noted in previous chapters, both Koenigs and Leau expressed a desire to find the maximal set of points which under iteration limit upon the fixed points or the periodic points of a given function. Koenigs in effect held up Cayley's success with Newton's method for the quadratic as a model to which studies of iteration might aspire [1884:s41]. Koenigs also observed, however, that while partitioning the plane into convergence regions is an important problem, it is a very difficult one, and he confessed that he had made virtually no progress on it [1884:s40].

In order to discuss Koenigs' attempt to divide the entire plane into regions according to their behavior under iteration of a given function $\phi(z)$, it is necessary to introduce some notation. Let $P = \{x_0, \ldots, x_{p-1}\}$ denote an attracting periodic orbit. If there is more than one periodic orbit, denote them by P_i. The maximal set of points on the Riemann sphere converging to P under iteration by $\phi(z)$ is denoted A_P. The set A_P is often called the *total domain of attraction* or the *basin of attraction* for the periodic orbit P. A point in A_P is said to be attracted to the orbit P.

[3] A point x_0 is *periodic with period* $p \geq 1$ if p is the smallest integer such that $\phi^p(x_0) = x_0$; such a point is also often referred to as a *period p point*. A fixed point then is a period 1 point. A periodic *orbit* is the set $\{x_0, \ldots, x_{p-1}\}$, where $x_i = \phi^i(x_0)$. The quantity $\frac{d}{dz}(\phi^p(x_i)) = M$ is independent of the choice of x_i, as can be easily verified via a routine calculation involving the chain rule. A periodic orbit P is termed attracting if $|M| < 1$; likewise, a point z is said to be attracted to a periodic orbit P if the limit points of the set $\{\phi^n(z) : n = 1, 2, \ldots\}$ equals the set P.

[4] A set is totally disconnected if its components are single points. It is perfect if it equals the set of its limit points.

The difficulties which Koenigs encountered in attempting to partition the sphere into domains of attraction A_{P_i} are twofold. First of all, determining the sets A_{P_i} in general requires sophisticated set theoretic techniques. It is no exaggeration to say that Koenigs, and those who followed him in the nineteenth century study of iteration, operated without a theory of sets. More will be said about this matter shortly, but in Koenigs' time at least, set theory was not a standard part of the French study of mathematics.

The second difficulty Koenigs encountered came with his observation that as n ranges from 1 to infinity, the set of equations

$$\phi^n(z) = z$$

may have infinitely many solutions, thus, the function $\phi(z)$ might well have infinitely many periodic orbits. Koenigs feared that the existence of infinitely many periodic orbits implied that there would in general be infinitely many *attracting* periodic orbits, hence the division of the plane into regions A_{P_i} would be quite complicated. On one hand, such fears were largely unfounded since both Fatou and Julia later showed that for a rational function $\phi(z)$, the number of attracting orbits is finite (see Corollary 11.10).

On the other hand, Koenigs was correct in assuming that the division of the Riemann sphere into regions of attraction might be quite complicated since Fatou and Julia showed as well that a set A_{P_i} can consist of infinitely many components. Koenigs was also correct in assuming that the problem of classifying regions of the sphere according to their behavior under iteration is an important one, since in many respects this could be regarded as the ultimate goal of the studies of Fatou and Julia, and of complex dynamics as a whole.

In illustrating the problems involved in delimiting the entire basin of attraction for a fixed point, I will assume that the periodic orbit P consists of a single fixed point x and will denote its basin of attraction by A_x. The ideas discussed below generalize to the case where P is an orbit of period p via consideration of the function $\phi^p(z)$.

Recall that both Schröder and Koenigs showed that if a given function $\phi(z)$ has an attracting fixed point x, then an open disc D can be found such that for all z in D, the orbit of z converges under iteration to x. One possible way to determine the basin of attraction A_x would be to consider the union of the preimages of D under $\phi^n(z)$, as n goes to infinity. Indeed, Fatou's 1906 note was based on such an approach.

However, the determination of the set A_x via this approach would have presented tremendous difficulties to Koenigs and his nineteenth century compatriots because of their unfamiliarity with set theory. Their work is devoid of all but the most rudimentary set theoretic or topological notions. There is little use of the concepts of open or closed sets, and the boundary of a set is never defined. The types of sets considered are rather tame by contemporary standards, and there is no mention whatsoever of nowhere dense or totally disconnected perfect sets. As will be indicated below, the latter occur naturally in the study of iteration. Finally, nowhere in the studies of Koenigs and his nineteenth century successors are infinite intersections or unions considered.

7.3 Fatou's Application of Set Theory

Fatou was evidently the first to apply set theoretic and topological techniques to the iteration of complex functions. It may seem surprising, in light of the fact that modern set theory began with Georg Cantor's work in the 1870's, that it took so long for someone to think to apply such notions to the study of iteration, but bear in mind that around the turn of the century, the study of the dynamics of complex functions was primarily a French activity, and the French as a whole were quite slow to appreciate Cantor's work.

This contrasted with the situation in other European countries, especially Italy and Germany, where Cantor's work was embraced and extended by many mathematicians in the 1870's and 1880's.[5] Yet, despite the appearance of French translations of several of Cantor's papers in *Acta Mathematica* in 1883, it was not until the 1890's that topology and set theory began to take root in France.

One of the earliest French examples of a set theoretic treatment of a mathematical idea was Camille Jordan's (1838–1922) concept of inner and outer content, which he presented in the first volume of the second edition of his *Cours d'analyse de l'École Polytechnique*, published in three volumes over the years 1893-96. As the following remark indicates, Jordan evidently helped familiarize Henri Lebesgue (1875–1941) as well as a number of other young French mathematicians, including Émile Borel (1871–1965) and René Baire (1874–1932), with various set theoretic notions:

> In having incorporated aspects of the theory of sets into his course at the École Polytechnique, Jordan, in a sense, rehabilitated this theory; he affirmed that it was a useful branch of mathematics. He in fact did more than affirm it, he proved it through his researches on integration and on the measure of areas and sets, which along with his studies on the rectification of curves, trigonometric series and *analysis situs*, prepared the way for certain works, mine in particular [Lebesgue 1922:16].

Borel, in his *Leçons sur la théorie des fonctions* published in 1898, was the first French mathematician to systematically apply Cantorian notions to the theory of complex functions. Indicative of the novelty of his approach is the following remark taken from the introduction:

[5] Among those influenced by Cantor's work were the Italians Ulisse Dini, Giulio Ascoli, Vito Voltera and Giuseppe Peano, the Swede Ivar Bendixson, and the Germans Paul du Bois-Reymond and Axel Harnack. See, for example, [Dini 1878], [Ascoli 1883], [Voltera 1881], [Peano 1887], [Bendixson 1883], [du Bois-Reymond 1882] and [Harnack 1881].

> ... it seems to me that it would be useful to expound upon, in an elementary way, certain relatively recent, but very important, advances. Among these is the theory of sets with which I begin this book [1898:vi–vii].

Borel's subsequent discussion involved such notions as the cardinality of infinite sets, derived sets, the Borel compactness theorem and the concept of the measure of sets. Moreover, it is clear from the way in which he discussed these matters that he felt he was expounding upon a subject with which his audience might not be very familiar.

Baire applied Cantorian notions to the study of real functions in three notes published in *Comptes rendus* [1897], [1898a] and [1898b], as well as in his thesis [1899]. Lebesgue, drawing in large part upon the work of Baire, Borel and Jordan, created his theory of integration and measure, which he presented in his thesis [1902], as well as in several notes in *Comptes rendus* between 1899 and 1901.

However, the conviction that rigorous topological and set theoretic notions were particularly useful tools in the study of analysis was not shared by all French mathematicians. Indeed, there was a certain reluctance among many of the old guard to embrace the new ideas contained in the works of Baire, Borel and Lebesgue. Reflective of this attitude is the famous remark Charles Hermite (1822–1901) made in a letter to Thomas Jan Stieltjes (1856–1894): "I turn away with fright and horror at the lamentable plague of functions without derivatives [Hermite-Stieltjes 1905,II:318]."

This antipathy towards many of the strange and intriguing ideas contained in the work of this new generation of French mathematicians is reflected in another oft-quoted remark:

> The horror manifested by Hermite is shared by nearly everyone, and I can scarcely take part in a mathematical conversation without encountering an analyst who says to me, "This will not interest you, it concerns functions which have a derivative," ... [Lebesgue 1922:99–100].

Fatou was one mathematician who did not share the horror of Hermite. He had the good fortune to be at the École Normale in the first years of the new century and was drawn to the new mathematical ideas being discussed there. Lebesgue supervised his doctoral thesis [1906b], and Fatou spoke warmly about his advice and encouragement [1906b:338]. The influence of Lebesgue is evident in Fatou's thesis, which treated the convergence of series from the point of view of Lebesgue's theory of integration, and included a proof of what is commonly referred to as Fatou's Lemma.

The fresh approach Fatou brought to the study of iteration is a favorable consequence of the movement led by Lebesgue, Borel and Baire, and serves as confirmation that so-called pathological objects studied by Lebesgue and others do occur in the normal course of events. Fatou's note [1906a] not only demonstrated that totally disconnected perfect sets occur naturally in the study of iteration, but in what was perhaps a rejoinder to the Hermitian point of view, he suggested that curves which lack tangents at infinitely many points do as well. This suggestion came in the form of an example, namely, the function

$$\phi(z) = \frac{z + z^2}{2},$$

which has attracting fixed points at 0 and ∞.

Fatou claimed that the basins of attraction for these two fixed points were separated by what he termed a non-analytic curve, by which he presumably meant a curve with no tangents on a dense set of points [1906a:548]. He claimed that this situation was by no means exceptional and that for many functions, basins of attraction were separated by such curves. He went on to state that the reasoning he provided for the special case of $\phi(z)$ could be applied to the general case. However, his argument for the special case was a bit unclear, and he gave no specific indication of what the general case might be. He may well have been aware of the defects of this early approach, because when he returned to this example near the end of World War I, he approached it from a completely different viewpoint, and gave sufficient conditions that the boundary curve of a basin of attraction be non-analytic, thereby providing a convincing argument that such curves were quite typical [1920b:240].

In light of the above discussion, Koenigs' evident unfamiliarity with set theory is to be expected. Nor should it seem too surprising that Leau did not employ set theory to his study [1897] despite the fact that he received his doctorate from the University of Paris in 1897, at a time when revolutionary developments in the subject were going on right around him. Rather, it should reinforce the fact that the study of set theoretic and topological notions was by no means a standard part of the French mathematical canon in the waning years of the nineteenth century.

7.4 Pierre Fatou

Born in 1878, Fatou died in 1929. He studied at the École Normale Supérieure from 1898 to 1901 and received his doctorate from the University of Paris in 1907. From 1901 until his death Fatou worked at the Paris Observatory and in 1928 was appointed its Titular Astronomer. According to Nathan, the reason Fatou worked at the Paris Observatory rather than at a university is simply that there were not many positions in mathematics available in Paris around the turn of the century [1980:547].

Despite his important contributions in several areas of mathematics, Fatou was not terribly prolific, especially in the early years of his career. Perhaps due to his responsibilities at the Paris Observatory, his mathematical output was rather meager for a number of years following his thesis. For the years 1907–1916, the listing for Fatou in Poggendorff [1922] cites only his thesis and two short papers on the convergence of series which apparently grew out of his thesis.

Fatou's *Comptes rendus* note on iteration [1906a] was the only work Fatou published on iteration until 1917. One wonders whether, as is the case with many notes which appeared in *Comptes rendus*, Fatou intended to quickly follow [1906a] with a research paper detailing and expanding upon the results sketched in his note.

Fatou's return to the subject of iteration near the end of World War I seemed to spark a creative burst. He published four notes in *Comptes rendus* on iteration in 1917–1918, and shortly thereafter published a major work on iteration, his *Sur les équations fonctionelles*, which appeared in three parts as [1919, 1920a, 1920b].[6] Fatou's mathematical output picked up considerably after the publication of these last papers, and in the 1920's he published several articles on a wide range of subjects including the theory of iteration, algebraic functions, multi-valued functions and astronomy.

7.5 Fatou's 1906 Note

Several factors suggest that Fatou saw his note [1906a] as a continuation of Koenigs' work in both iteration and functional equations. For example, in [1906a] Fatou refers to Koenigs' fixed point theorem, as well as his solution to the canonical Schröder equation

$$B(\phi(z)) = \phi'(0)B(z)$$

in the neighborhood of an attracting fixed point 0. Later, in his *Notice sur les travaux scientifiques*, Fatou discussed what he referred to as Koenigs' "remarkable" work, and observed that, subject to $\phi(z)$ satisfying certain hypotheses, he was able in [1906a] to extend Koenigs' local solution of the canonical Schröder equation to the entire plane, except for a totally disconnected perfect set [1929:17].

Fatou also observed in [1929] that Koenigs based the solution of several other functional equations on the Schröder equation, and that his own results extended to those functional equations as well. This echoes sentiments expressed in [1906a], where he remarked that his extension of Koenigs' solution to the Schröder equation applies to the Abel equation as well as to other "equations considered by Koenigs [1906a:547]."

Fatou's success with the Schröder equation in [1906a] turned upon his skills in set theory which gave him the means to describe the global behavior under iteration of a certain class of rational functions. Particularly useful was his familiarity with the technique of taking infinite intersections and his knowledge of totally disconnected

[6] This work was most likely Fatou's response to the French Academy of Sciences' announcement in 1915 that its 1918 *Grand Prix des Sciences mathématiques*, as well as 3000 francs, was to be awarded to the best work it received concerning the study of iteration. Julia was awarded the *Grand Prix*, but the Academy did award Fatou a prize of 2000 francs for his work in iteration.

perfect sets, henceforth abbreviated as TDP sets. The TDP sets which Fatou described in [1906a] are a complex analog of real, one dimension TDP sets which are often called Cantor sets. Cantor's middle-thirds construction was, however, by no means the first TDP set. The first such set appeared in Henry J. Smith's (1826–1883) paper [1875], and in the early 1880's, Paul du Bois-Reymond (1831–1889) in his paper [1880] and Vito Voltera (1860–1940) in his paper [1881], each apparently unaware of Smith's paper, constructed real TDP sets. It was not until 1883 that Cantor presented his famous middle-thirds construction.[7] Knowledge of these sets was slow to disseminate, and there seem to be few, if any, examples of TDP sets in French mathematics at the time of Koenigs' work in iteration. With the work of Lebesgue at the turn of the century, there were numerous examples of real, one dimensional TDP sets by the time Fatou's note on iteration appeared in 1906.

Two dimensional TDP sets were another matter altogether. Indicative of the fact that not much was known about two dimensional TDP sets in the first years of the twentieth century, is the following quotation from Ludvic Zoretti's entry on set theory in the 1912 French edition of the *Encyclopédie des sciences mathématiques pures et appliquées*:

> Totally disconnected sets have not been well studied until recently. ... Certain properties of these sets have a paradoxical allure. The most simple examples are furnished by one dimensional nowhere dense perfect sets. The two dimensional variety, however, constitute a very curious grouping of points [1912:142].

Fatou contributed greatly to the knowledge of two dimensional TDP sets in his note [1906a] where he showed that TDP sets turn up frequently in the study of the iteration of complex functions. Fatou claimed that for a class of rational functions typified by the family

$$\phi_k(z) = \frac{z^k}{z^k + 2},$$

the set of points J which iterate neither to an attracting fixed point nor to an attracting periodic orbit form a TDP set.[8] Fatou, in fact, sketched a proof of the following theorem, presented here in a somewhat more contemporary fashion than what is found in [1906a].

Theorem 7.1 (Fatou) *Let the function $\phi(z)$ be a rational function of degree strictly greater than* 1 *which has a unique attracting fixed point, which, without loss of generality, can be assumed to be* 0. *Suppose as well that there exists a neighborhood of* 0 *which contains all the critical values of $\phi(z)$ and on which all points converge to* 0 *under iteration.*[9] *Then the set of points J on the Riemann sphere which do not*

[7] The interest of Smith, du Bois-Reymond and Voltera in real, one-dimensional TDP sets was motivated by their researches in integration. See Hawkins [1975] for more detail. Cantor, who was aware of previous constructions of TDP sets, produced his middle-thirds set in [1883:590] as an example of a perfect set that was not an interval.

[8] For arbitrary functions, the set J is nowadays called the Julia set.

[9] A critical value is the image of a critical point.

converge to the fixed point 0 *under iteration by* $\phi(z)$ *is a TDP set.*

Sacrificing a bit of generality for clarity, I will apply the essence of Fatou's argument to an example he gave, namely, the function

$$\phi_2(z) = \frac{z^2}{z^2+2}.$$

This function has a unique attracting fixed point at 0, and all points in the open unit disc D converge to 0 under iteration by $\phi_2(z)$.[10] If an arbitrary point z in $\bar{\mathbb{C}}$ converges to 0 under iteration by $\phi_2(z)$, there exists a positive integer N such that for $n > N$, $\phi_2^n(z)$ is in D. Conversely, if a point z does not converge to 0 under iteration, there exists no $n \geq 0$ such that $\phi_2^n(z)$ is in D. In other words, there is a dichotomy between those points which converge to 0 and those which do not: a point z converges to 0 under iteration by $\phi_2(z)$ if and only if an iterate of z eventually lands in D. With this dichotomy in mind, Fatou constructed the set of points J which do not converge to 0 under iteration by $\phi_2(z)$ by finding all points z such that $\phi_2^n(z)$ is never in D.

In order to describe Fatou's construction of the set J, it will be convenient to denote the complement of the open unit disc D as E_1 and to let E_n equal the preimage of E_{n-1} under the function $\phi_2(z)$. Fatou claimed that

$$J = \bigcap_{n=1}^{\infty} E_n$$

and suggested a geometric construction of J. Since many of the TDP sets which occur in the study of complex dynamics can be constructed similarly, I will outline his construction for the special case of $\phi_2(z)$.

For $n > 1$, the sets E_n described above have as their boundary 2^{n-2} figure eights. For example, it may be shown that the preimage of the unit circle, denoted by E_1, is the hyperbola $y^2 - x^2 = 1$ where $z = x + yi$. This hyperbola is denoted Γ. Since the branches of Γ meet at ∞ it is helpful to view Γ as a figure eight on the Riemann sphere (see figure 7.1). The continuity of $\phi_2(z)$ implies in turn that the preimage of E_1, denoted by E_2, is the closure of the region interior to Γ, hence E_2 has the hyperbola Γ as its boundary.

Since $\phi_2(z)$ is a two-to-one map, all points on Γ have two preimages interior to Γ, hence the preimage of Γ consists of two figure eights, γ_1 and γ_2, each inside a distinct branch of Γ, as shown in figure 7.2. The set E_3 is the union of the closures of the interiors of the γ_i. Each γ_i has as its preimage two more figure eights, and the set E_4 is thus bounded by four figure eights.

Continuing in this manner it is apparent that the set J is the infinite intersection of the closures of the interiors of a nested sequence of figure eights. Although

[10] An easy way to see this is to first show that $\phi[D] \subset D$ and, since $\phi_2'(0) = 0 < 1$, apply the version of Schwarz's Lemma proved by Julia (see Theorem 6.2).

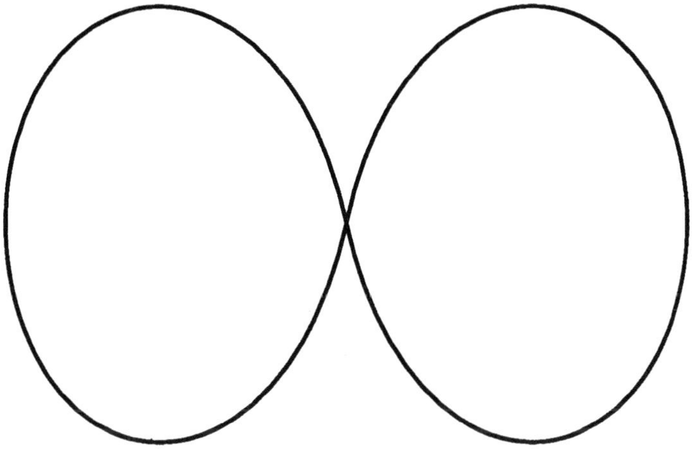

Figure 7.1: *The hyperbola $y^2 - x^2 = 1$ is denoted in the text as Γ. Infinity is the point where the two branches of Γ meet. The complement of the unit disc is E_1, and the union of Γ and its interior is E_2, therefore the figure eight Γ is the boundary of E_2. Points z such that $\phi_2^n(z)$ remains interior to Γ for all n form a TDP set.*

the details will be skipped, the diameters of these figure eights approach 0 as n approaches infinity, and the set J can therefore be shown to be a TDP set.

This construction is reminiscent of the construction of the Cantor middle-thirds set. Just as the Cantor set is what remains in $[0,1]$ after the successive removal of the intervals $(\frac{1}{3}, \frac{2}{3})$, $(\frac{1}{9}, \frac{2}{9})$, $(\frac{7}{9}, \frac{8}{9})$ and so forth, the set J is what is left in E_1 after the removal of the sets $E_m - E_{m+1}$ for $m \geq 1$.

Fatou's theorem is a significant result. Many functions behave under iteration as does the function $\phi_2(z)$. For example, if $|c| > 2$ the function $q_c(z) = z^2 + c$ satisfies the hypotheses of Fatou's theorem, although the fixed point is at ∞ not 0: its critical values are c and ∞, and it can be shown that for $|z| \geq |c| > 2$, $q_c^n(z)$ converges to ∞ (see, for example, [Devaney 1989:270]).[11] It is because Fatou's result describes a behavior shared by a class of functions more extensive than either the set of linear fractional transformations or the Newton's method function for an arbitrary quadratic that it is the first general result of a global nature.

The techniques Fatou used in the proof of the above theorem rely, however, on the fact that there is a unique attracting orbit, namely, the fixed point at 0, and therefore do not readily extend to a more general setting. In fact, the set J can only be a TDP set if there is a unique attracting orbit consisting of a fixed point.

[11] The function $\phi_2(z)$ behaves similarly to the quadratic $q_c(z) = z^2 + c$, where c is in the complement of the Mandelbrot set.

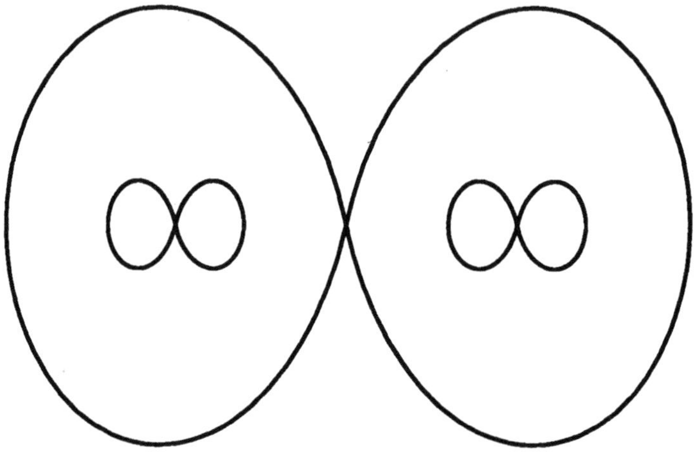

Figure 7.2: *The preimage of* Γ *under* $\phi_2(z)$ *consists of two figure eights,* γ_1 *and* γ_2*, one inside each branch of* Γ*, which form the boundary of* E_3*. The set* E_3 *is the union of closures of the regions interior to the* γ_i*. The region* $E_2 - E_3$ *is the area bounded by* Γ *and the* $gamma_i$*.*

If there is more than one attracting orbit then the preimages of a neighborhood D of a fixed point will instead limit on a closed curve whose structure can be quite complicated. Lest I leave the impression that Fatou gave no thought to this last situation in [1906a], the reader is reminded that, as was noted at the end of Section 7.2, Fatou closed his note with a brief and somewhat unsatisfactory discussion of an example of a function with two attracting fixed points.

In any event, due to the relative simplicity of the class of functions he considered, Fatou's exploration did not lead immediately to a global theory regarding the iteration of arbitrary rational functions. Such a development did not occur for another ten years, and it relied heavily on a body of theory that was only just beginning to make itself known, Montel's theory of normal families.

7.6 Fatou and Functional Equations

After sketching his proof of Theorem 7.1, Fatou related his work to Koenigs' study of functional equations. He observed that if in addition to satisfying the hypotheses of the above theorem, a function $\phi(z)$ also satisfies $0 < |\phi'(0)| < 1$, then the canonical Schröder equation

$$B(\phi(z)) = \phi'(0)B(z) \tag{7.1}$$

has a solution $B(z)$ which is analytic on $\bar{\mathbb{C}} - J$.[12]

Although Fatou did not justify this observation in his note [1906a], he indicated in his paper [1920b] a general method to extend the solutions of functional equations. Applied to the canonical Schröder equation, where $0 < |\phi'(0)| < 1$, this method is as follows:

Koenigs' solution $B(z)$ to the Schröder equation (7.1) is defined on a neighborhood D of the origin. Let $\tilde{D}$ be the preimage of D under $\phi(z)$. Define $B^*(\tilde{z})$ for $\tilde{z}$ on $\tilde{D}$ as

$$B^*(\tilde{z}) = \frac{1}{\phi'(0)} B(\phi(\tilde{z})).$$

This definition agrees with the original definition of $B(z)$ on D because if $\tilde{z}$ in $D \cap \tilde{D}$, then

$$B^*(\tilde{z}) = \frac{1}{\phi'(0)} B(\phi(\tilde{z})) = \frac{1}{\phi'(0)} \phi'(0) B(\tilde{z}) = B(\tilde{z}),$$

thus the notation $B^*(z)$ is unnecessary. Then since $\phi(\tilde{z})$ is in D,

$$\begin{aligned} B(\phi(\tilde{z})) &= (\phi'(0))^{-1} B(\phi(\phi(\tilde{z}))) \\ &= \frac{1}{\phi'(0)} \phi'(0) B(\phi(\tilde{z})) \\ &= \frac{1}{\phi'(0)} B(\tilde{z}). \end{aligned}$$

For the functions covered by Fatou's Theorem 7.1, the domain of definition of $B(z)$ can therefore be extended via a continuation of the above process throughout the domain of attraction of the fixed point 0, that is, throughout $\bar{\mathbb{C}} - J$, where J is a TDP set.

As Fatou noted during his discussion of Koenigs' work in [1929], Koenigs suggested in [1884] a similar means to extend his solution to the Schröder equation and remarked that his theorems—including his solution to the Schröder equation—can be extended to any disc $\tilde{D}$ in the set of points A which converge under iteration to the fixed point, but that "one knows nothing of the general manner in which this region $[A]$ is limited ... [1884:s40]." Without a theory of sets Koenigs would have had no idea just how far the preimages of D could have been extended.

Fatou's technique of extending the local solution of the canonical Schröder equation to a global solution foreshadows the strides he and Julia would make in their studies ten years hence when Montel's theory of normal families would allow a more general extension of Koenigs' local theory.

[12] It is interesting that, after noting that functions which satisfy the hypotheses of Theorem 7.1, as well as the condition $0 < |\phi'(0)| < 1$, also satisfy the Schröder equation (7.1), the only examples Fatou gave of functions which satisfy the hypotheses of Theorem 7.1 were the functions $\phi_n(z) = z^n/(z^n + 2)$. Since $\phi_n'(0) = 0$, the Schröder equation at (7.1) has no relevance for these functions.

Chapter 8

Montel's Theory of Normal Families

8.1 Introduction

The key to understanding the behavior under iteration of an arbitrary point in the complex plane lies in understanding the set of points whose orbits do not converge to an attracting or neutral orbit. Fatou's note [1906a] described this set, often denoted J, in detail for a class of complex rational functions possessing a unique attracting fixed point. Although his technique of examining the intersection of the preimages under $\phi^n(z)$ of the complement of a neighborhood of an attracting fixed point led to his discovery that when $\phi(z)$ has a unique attracting fixed point the set J can be a totally disconnected perfect set, this technique did not reveal enough about J when $\phi(z)$ has more than one attracting orbit.

Another possible way of investigating J would be via direct calculation, but before the advent of the computer, the difficulties in this approach were imposing, since for large n and arbitrary z the calculations involved in directly computing $\phi^n(z)$ are daunting. Given this resistance to frontal assault, some sort of heavy machinery was called for. Fortunately, it was provided by a young French mathematician, Paul Montel, who, during the fallow period in the development of complex dynamics which followed Fatou's 1906 note put the finishing touches on his theory of normal families.

A normal family of complex analytic functions may be defined as follows:

Definition 8.1 *A family of functions $\mathcal{F}$ which is analytic in the interior of a domain D is normal in D if all infinite sequences of functions from $\mathcal{F}$ contain a* *subsequence which converges uniformly on all compact sets D' interior to D.*[1] *This limit function is either analytic or identically infinite.*[2] *Normal families of meromorphic functions are defined analogously, as are families of real functions.*

Montel's theory of normal families was quite powerful. In the first two decades of the twentieth century Montel applied his theory of normal families to a variety of subjects within complex function theory. He did not, however, apply this theory to the study of complex iteration—that was first done independently by Fatou and Julia around 1917. Montel perhaps wished that he had done so, for not only did the theory of normal families have a profound effect on the study of iteration, but Montel himself seemed quite taken with the role that his theory played in the studies of Fatou and Julia. In his 1927 book *Familles Normales*, he made the following comment about the use of his theories in what he termed the "remarkable works" of Fatou and Julia:

> One of the most important applications of the theory of normal families is found in the study of iteration of analytic functions and the solutions of the functional equations to which they are related [1927:213].[3]

After showing that for an arbitrary complex rational function $\phi(z)$ the family of functions

$$\{\phi^n\},$$

where n is a non-negative integer, is generally not normal on neighborhoods of points from the set J described above, Fatou and Julia were able to prove several important results regarding the iteration of complex functions. Their use of Montel's theory of normal families is important for another reason as well, since it also represents the introduction of sophisticated theorems from complex function theory into complex dynamics. Prior to their work, the study of complex iteration stood curiously apart from complex function theory in the sense that it relied principally on results developed internally, such as fixed point theorems and theorems pertaining to the solution of various functional equations. Exceptions to this were Schröder's initial analysis of Newton's method from the point of view of elementary complex function theory, and Fatou's application of set theory in 1906.

Before beginning my discussion of the work of Fatou and Julia in Chapter 11, I will describe the events and ideas which led to the development of Montel's theory of normal families, and briefly outline the subjects to which he applied his theory.

---

[1] A domain is an open, connected subset of the complex plane. A function is meromorphic on D if it has a pole on D.

[2] A sequence $\{f_n\}$ converges uniformly to infinity on a domain D if for any $\epsilon > 0$ there exists an N such that $|f_n| > \epsilon$ for $n > N$ on any compact set D' interior to D.

[3] Montel's treatment of the theory of iteration in this book is self-contained and serves as an excellent introduction to the subject.

8.2 Paul Montel and Normal Families

Paul Montel was born in Nice in 1876 and died in Paris 99 years later. He completed his studies at the École Normale Supérieure in 1897 and evidently did not commence work on his doctorate for several years. Instead, he divided his time between travel, literature and the teaching of mathematics at various provincial secondary schools. His friends apparently felt that he was wasting his considerable mathematical gifts and urged him to return to the study of mathematics [Dieudonné 1990:649]. Apparently they prevailed, and he finally turned his attention full time to mathematics. In 1907 he received his doctorate from the University of Paris, although notes concerning topics he treated in his thesis appeared in the *Comptes rendus* of the French Academy of Sciences as early as 1903.

Like Fatou, Montel was intrigued by the new generation of French mathematicians, including Baire, Borel and Lebesgue, and again like Fatou, Montel completed his dissertation [1907] under Lebesgue's direction. His dissertation concerned infinite sequences of both real and complex functions, and although he did not explicitly use the term "normal families" he used the concept throughout his thesis. After earning his doctorate, he taught at various secondary schools including the Lycée Buffon in Paris.

Montel's mathematical prowess quickly brought him renown. He published two mathematical texts in 1910, *Algebra*, co-authored by Borel, and *Leçons sur les séries de polynomes*, which was part of the *Collection de monographs sur la théorie des fonctions* published under the direction of Borel. He became *maître de conferences* at the University of Paris in 1911, and, in 1913 his work in the theory of analytic functions won him the French Academy of Sciences' *Prix Gustave Roux*, a prize whose purpose was to honor young French mathematicians of exceptional talent. In 1918 the Academy awarded him a second prize, its *Prix Francœur* for his work concerning "sets of analytic functions." Montel was promoted in 1922 to full professor at the University of Paris. During World War II he served as Dean of the Faculty of Science, where, according to Dieudonné, he upheld the dignity of the university in spite of the German occupation.

He retired in 1946 and was subsequently honored with several honorary positions and degrees, both in France and abroad. He was elected to the French Academy of Sciences in 1937 and was named its president in 1958. Throughout his career Montel evidently aided in the development of many French mathematicians, and in his necrology of Montel, Szolem Mandelbrojt observed that "at least a quarter of our present *confrères*" had Montel as *rapporteur* of their thesis [1975:186].

Montel worked principally in complex function theory, and it is his theory of normal families for which he is most famous. Although Montel was the first to develop a cohesive theory around the concept of a normal family, there had been interest in such families long before Montel. What marked Montel's use of normal families was his application of the concept in areas where it had not been used previously. In particular he extended the notion of a normal family to complex function theory where he applied it with great success to such topics as convergence of sequences and series, Picard theory and conformal mapping.[4]

Montel's first mature works on normal families are the papers [1912], [1916] and [1917]. The results from these papers are also presented in his book *Familles normales*, first published in 1927. In both [1912] and [1916] he treated Picard theory, and in [1917] he proved the Riemann mapping theorem as well as several other theorems related to conformal mapping.

8.3 The Influence of Ascoli and Arzelà

Although Montel was the first to use the phrase "normal family," the concept was in use well before Montel defined it. In an attempt to avoid anachronistic use of the term, the phrase "Property N" will be used to indicate the use of the concept of normality prior to Montel's first use of the term. The nineteenth century conception of Property N was a little different than the definition of normality given at the beginning of this chapter since the limit function usually was not allowed to be infinite.

Principal among those who studied families with Property N in the nineteenth century were Cesare Ascoli (1843–1896) and Giulio Arzelà (1847–1912). Each of these mathematicians offered early proofs of what is nowadays a central theorem in the study of normal families, the so-called Ascoli-Arzelà Theorem, which is stated below for complex functions of a single variable.

Theorem 8.2 (Ascoli-Arzelà Theorem) *Let $\mathcal{F}$ be an equicontinuous family of complex functions defined on a domain D with the property that for each $z \in D$ the set $\{f : f \in \mathcal{F}\}$ is bounded.*[5] *Then $\mathcal{F}$ is a normal family.*

It should be emphasized that Ascoli and Arzelà each proved this theorem only for real functions; in addition, both mathematicians were apparently working under the implicit assumption that the families of functions they considered were uniformly bounded.

The interest of both Ascoli and Arzelà in the above theorem was evidently stimulated by the Dirichlet principle which in turn relates to the Dirichlet problem, named after the German mathematician Peter Gustave Lejeune-Dirichlet (1805–1859). The Dirichlet problem had its origins in the study of potential theory and was an extremely important problem throughout the nineteenth century. The Dirichlet

[4]The so-called Big and Little Picard Theorems, first proved by Émile Picard (1856–1941) in 1879, spawned a number of papers beginning with Borel's paper [1896]. This body of work is sometimes collectively referred to as Picard theory. A discussion of the role of Picard theory in the development of Montel's theory of normal families will be given below in section 8.7.

[5]The family $\mathcal{F}$ is equicontinuous on D if for every $\epsilon > 0$ there exists a corresponding $\delta > 0$ such that for $f(z) \in \mathcal{F}$ and for any x, y in D, $|x - y| < \delta$ implies that $|f(x) - f(y)| < \epsilon$.

problem is stated below for $\mathbb{R}^n$, but in the 1800's interest was centered in the $\mathbb{R}^2$ and $\mathbb{R}^3$ cases:

Dirichlet Problem *Let D be an open set in $\mathbb{R}^n$. Let $f(x)$ be a continuous real function which is defined on ∂D. Find a function $F(x)$ which is harmonic on D and which equals $f(x)$ on ∂D.*[6]

The Dirichlet problem was considered by many nineteenth century mathematicians including Dirichlet himself, Carl Gauss (1777–1855) and George Green (1793–1841).[7] In the first two-thirds of the nineteenth century, solutions to the Dirichlet problem generally relied on the Dirichlet principle:

Dirichlet Principle *Among all the real-valued functions $f(x)$ which are C^2 on D and which can be extended continuously to ∂D, there exists a set of functions $\mathcal{A}$, often called* the set of admissible functions, *which contains a function $F(x)$ minimizing the* Dirichlet integral

$$DI(f) = \int_D \left[\left(\frac{\partial f}{\partial x_1}\right)^2 + \cdots + \left(\frac{\partial f}{\partial x_n}\right)^2\right] dx, \tag{8.1}$$

where $f(x) \in \mathcal{A}$. Moreover, the minimizing function $F(x)$ satisfies $DI(F) = 0$.

In $\mathbb{R}^2$ the Dirichlet principle was extremely important to the study of complex analysis since Riemann based his proof of the Riemann Mapping Theorem, and consequently his theory of complex functions, on his acceptance without proof of the Dirichlet principle. Such faith in the Dirichlet principle, which evidently rested on intuition, was the norm among mathematicians well into the second half of the nineteenth century and is reflective of a widespread lack of concern regarding existence proofs. The Dirichlet principle is tantamount to the assumption that if

$$I(f) = \int_D f(x)dx$$

has a greatest lower bound L for a particular admissible set of functions $\mathcal{A}$, then there exists a function F in $\mathcal{A}$ satisfying $I(F) = L$.

In 1870 Weierstrass offered the following example which undermined confidence in the Dirichlet principle, and hence in the Dirichlet problem itself:

Example 8.3 (Weierstrass) *Let $\mathcal{A}$ be the set of C^1 real functions defined on the interval $[-1, 1]$ such that $f(-1) \neq f(1)$. The infimum taken over $\mathcal{A}$ of*

$$\int_{-1}^{1} \left(x\frac{df}{dx}\right)^2 dx$$

[6]A C^2 function $F(z)$ is harmonic on D if $\sum_1^n \frac{\partial^2 F}{\partial x_i^2} \equiv 0$ on D.

[7]For a lively and more detailed treatment of the Dirichlet problem see Monna's book *Dirichlet's Principle*.

is the function $f(x) \equiv 0$, yet the zero function is not in $\mathcal{A}$.

Weierstrass thus showed that the existence of the $\inf I(f)$ taken over a particular admissible set $\mathcal{A}$ does not guarantee the existence of a function $F(x)$ from $\mathcal{A}$ satisfying $I(F) = \inf I(f)$.

Various attempts were made throughout the remainder of the century to solve the Dirichlet problem. Both Hermann Amandus Schwarz (1843–1921), in 1870, and Poincaré, in the late 1880's, solved versions of the Dirichlet problem which did not rely on the Dirichlet principle. In 1900 David Hilbert (1862–1943) rehabilitated the Dirichlet principle by proving it in the $\mathbb{R}^2$ case, subject to certain conditions on the boundary of D, the function $f(x)$ and the set of admissible functions $\mathcal{A}$. In the 1930's the Dirichlet principle was proved for $\mathbb{R}^n$ under quite general conditions via theorems from functional analysis.

The relevance of the Ascoli-Arzelà Theorem, stated above as Theorem 8.2, to the Dirichlet principle is evident: if the set of admissible functions $\mathcal{A}$ is equicontinuous, and if there exists a sequence of functions $\{f_n\}$ from $\mathcal{A}$ such that $\mathrm{DI}(f_n)$ converges to 0, then there exists some subsequence of $\{f_n\}$ which converges uniformly to a function $F(x)$ satisfying $\mathrm{DI}(F) = 0$.

8.4 Montel's Early Work

Montel's first published version of the Ascoli-Arzelà Theorem appeared in his *Comptes rendus* note [1904] where he claimed that uniform boundedness was a sufficient condition for Property N to hold. As indicated in the above discussion there was quite a bit of use of Property N in the theory of real functions long before 1904. Indeed, as the following quotation suggests, Montel's thesis [1907] was evidently stimulated by his observation that a thorough investigation into Property N was warranted since this property had proven so useful in a variety of situations:

> ... a great number of properties pertaining to the theory of functions are found in an equivalent form in the functional calculus. The demonstrations of these theorems concerning continuity, maximums, and minimums, demand only that the set $\mathcal{E}$ and the law of correspondence [i.e., the functional] are subject to certain conditions: 1) that from all infinite sequences one can extract at least one other infinite sequence having a limit element; 2) that these limit elements belong to E, in other words, that the set is closed. For the law of correspondence it is necessary, in general, to suppose that it is continuous.
>
> The principal difficulty in approaching the functional calculus from this point of view lies in verifying that the sets $\mathcal{E}$ have the properties mentioned above: I cite in this instance the works of Arzelà on the series of functions, that of Hilbert on the Dirichlet problem treated in the

> manner of Riemann, and that of Lebesgue on the problem of Plateau.[8] It seems thus desirable to extend our knowledge of sets of functions possessing the properties required for the set $\mathcal{E}$.
>
> In this work I have assayed to study two families of continuous functions, those which satisfy the following conditions: families of equicontinuous real functions, and families of [uniformly] bounded complex functions [1907:1–2].

Despite his mention of the Dirichlet problem, the problem of Plateau and Maurice Fréchet's (1865–1963) innovative study of function spaces, Montel did not focus upon these subjects in his thesis [1907]. Rather, he concentrated his efforts on the theory of real and complex functions.

The time between Montel's thesis [1907] and his book [1910], in many senses comprises the developmental period of his theory of normal families. In comparing these early works to [1912] and [1916] one gets the impression that in both [1907] and [1910] he was still in the process of sorting out the components of his theory of normal families. Suggestive of this, is the fact that he did not begin to use the term normal families until his paper [1912]. Although he used Property N to study complex function theory in [1907], something which had not been done previously, his application of Property N tended to be in areas where its efficacy had already been established in the theory of real functions.[9]

For example, one of the primary applications of Property N in both [1907] and [1910] was to the convergence of series and sequences of functions, an area in which Arzelà in his paper [1899] had already profitably used Property N. Although Montel's extension of Property N to complex families in [1907] and [1910] was certainly novel, he did not yet apply it to either Picard theory or conformal mappings.

In [1907] Montel also offered proofs of the Ascoli-Arzelà Theorem for both real and complex functions, which he then used to prove what is often called Montel's Theorem. Although he proved this theorem in [1907], the following statement is from [1910:21]:

Theorem 8.4 (Montel's Theorem) *If a family of complex analytic functions $\mathcal{F}$ is uniformly bounded on a domain D, then "from any infinite sequence of functions from the family, a new sequence can be extracted which converges uniformly in the domain D to an analytic limit function."*

The first step of Montel's proof of this theorem involved a very simple application of the Cauchy integral theorem which proves that uniformly bounded families of

---

[8] The problem of Plateau involved finding a minimal surface bounded by a given Jordan curve in $\mathbb{R}^3$. Like the Dirichlet principle, the problem of Plateau involves the problem of determining whether or not a function exists which minimizes a particular integral. The problem was finally solved in the 1930's.

[9] Contemporaneous with Montel's extension of Property N to complex analysis, Fréchet used Property N to study spaces of functions. As an example of his ideas, he applied Property N to sets of complex polynomials in his paper [1906]. Whether or not Montel influenced Fréchet, or vice versa, is uncertain, in part because Montel and Fréchet used Property N to different ends.

analytic functions are equicontinuous. His theorem follows immediately from the Ascoli-Arzelà Theorem.

8.5 Montel's Study of Convergence Issues

Montel's investigations into the convergence properties of sequences and series of analytic functions, which Julia and Fatou both used in their study of domains of attraction, relied in large measure on the Ascoli-Arzelà Theorem. Montel's interest was evidently motivated by his desire to find sufficient conditions that a series or sequence of analytic functions is analytic. In [1907] he asked, "In what case can it be assumed that the limit of a sequence of functions be an analytic function [1907:98]?" Later he remarked that "a problem which is linked inextricably to [series of analytic functions] is the search for the conditions which a function $F(z)$ must satisfy if it is the sum of a series of analytic functions [1907:100]."

This interest in convergence issues reflected a general interest among mathematicians of the time, in particular those involved in integration theory. Through the efforts of Weierstrass, Darboux, Koenigs and others, the usefulness of uniform convergence was well-established. It was, for example, well-known that a uniformly convergent sequence or series of analytic functions converged to an analytic limit, and that a sufficient condition that

$$\int_a^b \lim_{n\to\infty} \sum f_n = \lim_{n\to\infty} \int_a^b \sum f_n, \tag{8.2}$$

where the f_n are real-valued functions, was the uniform convergence of the series $\sum f_n$. There were therefore attempts to find other sufficient conditions that equation (8.2) hold or that the limit of a series or sequence of analytic functions was analytic. For example, in 1902 the American mathematician William Osgood (1863–1943) proved the following theorem [1902:25ff]:

Theorem 8.5 (Osgood) *Let $\{f_n\}$ be a sequence of complex analytic functions defined on a domain D, and let $S_n(z) = \sum f_n$. Suppose that $\{S_n\}$ is uniformly bounded on D and that the series $\sum f_n$ converges pointwise on a dense subset T of D. Then $\sum f_n$ converges on D to an analytic function.*

Osgood did not apply Property N in his proof; rather, he showed that the series $\sum u_n(x,y)$ and $\sum v_n(x,y)$ converge uniformly on D where $f_n(z) = u_n(x,y) + v_n(x,y)i$. At the conclusion of his proof he remarked that his theorem generalizes a result of Thomas Jan Stieltjes, who in his paper [1894] showed that if a series of complex analytic functions with uniformly bounded partial sums converges uniformly on a subdomain D' of D, it converges uniformly on the entire domain D [Stieltjes 1894:451ff]. Osgood's result shows that Stieltjes' condition of uniform convergence on the sub-domain D' could be replaced by pointwise convergence on a dense subset T of D'.

Montel cited Stieltjes' and Osgood's results and generalized them by showing that Osgood's condition that the subset T be dense in D can be replaced by the weaker condition that T be an infinite subset of D with a limit point in D [1907:76-77]. In his paper [1912] he generalized Osgood's theorem again by the following result, a result used later by both Fatou and Julia [1912:531]:

Theorem 8.6 (Montel) *Let $\{f_n\}$ be a sequence of functions from a meromorphic family $\mathcal{F}$ which is normal on a domain A. Suppose that the f_n converge to a function $G(z)$ on an infinite set of points which has a limit point in A. Then the f_n converge uniformly to $G(z)$ on A.*

8.6 An Important Result from [1912]

In his papers [1912] and [1916] Montel focused exclusively on normal families of complex analytic functions. He explicitly defined the concept of normality by name and precisely delineated the interrelationships among normal families, equicontinuous families and bounded families. He in addition proved several theorems concerning normal families, some of which had their genesis in [1907] and [1910]. For example, in [1912] he proved that if a family of analytic functions defined on a domain D is bounded away from a point α in the complex plane then it is normal.[10] He proved an identical result in both [1907] and [1910], but stated only that such families were equicontinuous.

One of the most important results from [1912] is the so-called Montel normality criterion which Fatou and Julia both used extensively in their studies of complex iteration. This result involves what is called an exceptional value for a family $\mathcal{F}$.

Definition 8.7 *A complex quantity w_0 is an* exceptional value *of a family $\mathcal{F}$ of analytic or meromorphic functions defined on a domain D if*

$$w_0 \notin \bigcup_{f \in \mathcal{F}} f[D].$$

Using this definition Montel proved the following:

Theorem 8.8 (Montel's Normality Criterion) *Let $\mathcal{F}$ be a family of analytic functions on D. If $\mathcal{F}$ has at least two exceptional values on D then it is normal. If $\mathcal{F}$ instead consists of meromorphic functions, then a sufficient criterion of its normality is that there be at least three exceptional values.*

[10] A family $\mathcal{F}$ defined on a domain D is bounded away from a point α if there exists an $\epsilon > 0$ such that for all $f \in \mathcal{E}$ and all $z \in D$, $|f(z) - \alpha| > \epsilon$.

By considering the family

$$\mathcal{G} = \left\{ g : g(z) = \frac{f(z) - a}{b - a} \text{ and } f \in \mathcal{F} \right\},$$

it can be assumed without loss of generality in the analytic case that the exceptional values a, b are $0, 1$ since if $f(z)$ can not take on a, then $g(z)$ can not equal 0; analogously, if $f(z)$ has b for an exceptional value, then 1 is an exceptional value of $g(z)$.

As will be indicated, Theorem 8.8 was crucial to Montel's study of the body of work stemming from the Picard Theorems. In [1912] he gave a very involved proof of Theorem 8.8 using a modular function from elliptic function theory, and then sketched a simple elementary proof of Theorem 8.8, that is, a proof which did not use a modular function, but was instead based on the following theorem proved by Friedrich Schottky (1851-1935) [1904:1255ff] and later modified by Edmund Landau (1877-1938) [1906:265–67]:

Theorem 8.9 (Schottky) *If $f(z)$ is an analytic function on a closed disc D centered at z_0 with exceptional values 0 and 1, then $f(z)$ is bounded on D by a constant M which depends only on α, where*

$$\alpha = \min \left\{ |f(z_0)|, \frac{1}{|f(z_0)|}, \frac{1}{|f(z_0) - 1|} \right\}.$$

This remarkable theorem implies that the bound on a function with two exceptional values on a closed disc depends only on the value of the function at the center and is quite useful in the study of Picard's Theorems.

Although Montel's normality criterion, Theorem 8.8, was not presented as a theorem until [1912], Montel was aware of it as early as 1907 as is indicated by his observation in a footnote included in [1907] that Schottky's Theorem can be used to show that "a family of functions which takes, in a domain D, neither the value 0 or 1 is an equicontinuous family [1907:74]." Although the proof is not too difficult, Montel gave no proof, nor did he make any use of this result in [1907].

8.7 Applications of Montel's Theory to Picard Theory

Judging by the amount of space he devoted to the body of work stemming from the theorems of Picard in both [1912] and [1916], Montel considered his study of Picard theory to be one of the most important applications of his theory of normal families. Previous to the appearance of Montel's paper [1912], the Picard Theorems had generated quite a number of papers by prominent mathematicians such as Borel,

Schottky, Landau, Paul Lévy, Constantin Carathéodory (1873–1950), and Ernst Lindelöf (1870–1946).[11]

The Little Picard Theorem states that any non-constant, entire complex analytic function must take on all values of the complex plane with the possible exception of one. Picard presented this result in his note [1879a] and proved it using a modular function. Picard's statement and proof of the Big Picard Theorem appeared shortly thereafter in his note [1879c]. The Big Picard Theorem states that on a neighborhood of an isolated essential singularity, an analytic function takes on all values of the plane with the exception of at most one. Picard's proof of this theorem also relied on a modular function.

A good deal of the early work in Picard theory was devoted to either the generalization of Picard's Theorems or the discovery of elementary proofs for them, that is, proofs which do not rely on a modular function or elliptic function theory, and Schottky's Theorem is useful in both of these endeavors. Borel gave the first elementary proof of the Little Picard Theorem in his note [1896]. In his paper [1904] Schottky drew on Borel's proof and used Theorem 8.9 to produce the first elementary proof of the Big Picard Theorem. Landau in his paper [1904] independently proved a less general version of Schottky's Theorem and modified it again in his paper [1906] while summarizing the existing literature concerning the Picard Theorems. Lindelöf simplified Schottky's elementary proof of the Big Picard Theorem in his paper [1909] and proved several new theorems concerning the behavior of functions in neighborhoods of isolated essential singularities. He also presented a paper on Picard theory, the paper [1910], to the 1909 International Congress of Mathematicians.

A common thread in the Picard Theorems, as well as in the subsequent theorems they generated, is an interest in analytic functions $f(z)$ which have two exceptional values on a disc D centered at the origin. Viewed in this light the Little Picard Theorem states that for any non-constant entire function $f(z)$ there is an R for which no disc D of radius greater than R has two exceptional values. The Big Picard Theorem states that there is no disc surrounding an isolated essential singularity on which $f(z)$ has two exceptional values. Typical of the results concerning exceptional values on a disc of radius R is Schottky's Theorem.

In addition to its importance to Picard theory, Schottky's Theorem was also very important to Montel. In [1912] he remarked that his normality criterion, Theorem 8.8, is a simple consequence of Schottky's Theorem and then proceeded to give new proofs of both Picard Theorems using his normality criterion. This suggested that the Picard Theorems might actually be a consequence of the theory of normal families. However, in the paper [1912] it was only a suggestion, since Montel needed to go beyond the theory of normal families and appeal to either elliptic function theory or Schottky's Theorem to prove his normality criterion.

All this was to change with Montel's paper [1916] in which he used the theory

[11] For a nice discussion of the mathematics involved in this body of work, see Chapter II of Sanford Segal's *Nine Introductions in Complex Analysis*.

of normal families to prove Schottky's Theorem, thus showing that his normality criterion, and all that followed from it, including the Picard Theorems, are in fact consequences of the theory of normal families. In addition, Montel's proof of Schottky's Theorem implies that any result which followed from Schottky's Theorem is in turn implied by the theory of normal families. This had tremendous implications to Picard theory since many results therein, including the Picard Theorems, formerly viewed as consequences of Schottky's Theorem, were actually revealed to be consequences of Montel's theory of normal families. Thus, not only did Montel provide a new elementary proof of the Picard Theorems, he also effectively embedded Picard theory in the theory of normal families.

Chapter 9

The Contest

9.1 Overview

In 1915 the French Academy of Sciences announced that it would award its 1918 *Grand Prix des Sciences mathématiques*—and 3000 francs—for the study of iteration. This announcement evidently motivated both Julia's *Mémoire sur l'iteration des fonctions rationnelles*, referred to as [1918], and Fatou's *Sur les équations fonctionnelles*, published in three parts as [1919], [1920a] and [1920b], as well as a third effort by Samuel Lattès.

9.2 Biographical Sketches of Lattès and Julia

Lattès was born in Nice in 1873 and died of typhoid fever in the summer of 1918. He was a professor at the University of Toulouse for most of his mathematical career and published several papers on function theory and the iteration of multi-variable functions. Lattès' entry to the contest was never published but some indication of its concerns can be gleaned from three notes which appeared in the *Comptes rendus* of the Academy of Sciences in 1918, the notes [1918a], [1918b] and [1918c], as well as from a summary of the results of the contest which appeared in the *Compte rendu* of December 2, 1918. Although Lattès' entry included a discussion of the iteration of rational functions of one complex variable, its principal interest was the iteration of meromorphic functions of two complex variables, a subject he had discussed earlier in several papers, including his papers [1907] and [1908]. Lattès' work differed significantly from that of Fatou and Julia in two respects. Not only did Fatou and Julia focus exclusively on the iteration of functions of a single complex variable, but, unlike Lattès, their approaches were rooted in Montel's theory of normal families. Because the concern of this book is the work of Fatou and Julia, I will not discuss Lattès' work involving the iteration of multi-variable functions. I will, however, discuss Lattès' approach to the iteration of single variable functions in Chapter 10.

Julia was born in 1893 in Algeria and died in Paris in 1978. Although his father was an uneducated craftsman, Julia's mathematical gifts were in evidence at a young age. He was educated by friars, and in 1911 placed first on the entrance exams to both the École Polytechnique and the École Normale Supérieure, as did Gaston Darboux in 1861, and, like Darboux, entered the latter. In 1913 Julia published his first mathematical paper in the *Bulletin de la Societe mathématique*.

Julia served in the French military as a sub-lieutenant during World War I, and in 1915, in the face of a furious German attack, he exhibited uncommon bravery and was awarded the French Legion of Honor. The price of Julia's bravery was extreme: he suffered a terrible wound to his face and was left with a disfigured nose which in future years he customarily covered with a patch. According to his medal citation, his wounds left him almost blinded and unable to speak, yet he issued a written directive that he was not to be evacuated until after the German attack was repulsed.

During his recuperation he was buoyed by visits from both George Humbert (1859–1921), from whom more will be heard shortly, and Émile Picard, as well as by the ministrations of his future wife, the daughter of the composer Ernest Chausson [Garnier 1978].

Julia quickly established himself as a prominent mathematician. He received his doctorate from the University of Paris in 1917, and soon afterwards won a major mathematical prize, the French Academy's *Prix Bordin* in 1917. In addition to his work involving the iteration of complex functions, Julia also made important contributions in number theory, analysis and the theory of Hilbert operators. Julia was honored with membership in the French Academy of Sciences in 1934 at the age of 41.

9.3 The 1918 *Grand Prix*

Mathematical contests such as the one soliciting works on iteration were quite common throughout the nineteenth century and into the early twentieth century, as was the awarding of prizes to worthy mathematical endeavors. Each year the Academy presented several prizes in mathematics, each accompanied by monetary awards. For example, in 1918 the French Academy awarded prizes worth at least 1000 francs to Fatou, Julia, Montel and Lattès.

The full text of the Academy's announcement that it would award its 1918 *Grand Prix des Sciences mathématiques* for the study of iteration is as follows:

> The iteration of a substitution of one or many variables, that is to

> say, the construction of a system of successive points $P_1, P_2, \ldots P_n, \ldots$, where each is deduced from the preceding by a given operation
>
> $$P_n = \phi(P_{n-1}) \; (n = 1, 2, \ldots, \infty)$$
>
> (where ϕ depends rationally, for example, on the point P_{n-1}) and where the initial point P_0 is also given, rises in many classical theories as well as in some of the most celebrated memoirs of Poincaré.
>
> Until now, the most well known works have been devoted totally to the "local" point of view.
>
> The Academy wishes that the study pass to the examination of the entire domain of the values which the variable can take on. In this spirit, it has, for the year 1918, put forth the following question:
>
> *To perfect in a meaningful way the study of the successive powers of a given substitution, when the exponent of the power grows indefinitely.*
>
> *One should consider the influence of the initial value P_0, the given substitution, and one may limit oneself to the most simple cases, such as the rational substitution of one variable* [*Comptes rendus*, December 27, 1915:921].

The above quotation indicates that the Academy felt that the time was ripe to stimulate the global study of iteration, and the reference to Henri Poincaré certainly suggests that his studies in celestial mechanics may have had something to do with the Academy's decision to offer the prize. However, it is doubtful that Poincaré's work was the primary motivating factor.

Poincaré did indeed iterate real variable functions with his so-called return map which he used to study the solutions to differential equations involving the planar, restricted three body problem, in which the largest body, for example the sun, is fixed and the effect of the second body on the motion of a third, smaller body is studied. He related the stability of the third body to the character of the set of limit points of the forward orbit of certain points under iteration by the return map. Poincaré was an extremely influential mathematician and had been a member of the Academy until his death in 1912. It is therefore possible that the Academy thought that a general study of the kind it suggested might shed further light on Poincaré's work.

Yet while Poincaré's influence may have played a role in the Academy's decision there were also several other compelling reasons for the Academy to consider the question of iteration of complex functions. As has already been documented, the iteration of complex functions was a well-studied subject in France, especially in the late nineteenth century. By 1915 there was a coherent corpus regarding the local behavior of the iterates of a complex function, as well as the beginning of a global study. The chief figure in the field, Gabriel Koenigs, was by 1915 a prominent French mathematician and scientist. He was a professor at the Sorbonne with scores of publications to his credit and had founded an important laboratory which played a major role in the French war effort. He was, in addition, on familiar terms with many members of the Academy, including Paul Appell and Émile Picard.

The study of iteration of complex functions had also been applied to other fields. As noted in Chapter 4.3, Appell in his paper [1891] wrote a solution to the Hill differential equation (4.1) which relied heavily on Koenigs' paper [1884]. Picard had also shown the utility of iteration in the parameterization of algebraic surfaces in his paper [1900], when he in effect iterated a complex function until it converged to an attracting fixed point [1900:21]. The iteration of complex functions also played an important role in the study of functional equations, a fact which has been discussed several times already. Salvatore Pincherle (1853–1936) devoted a significant portion of his discussion of functional equations in both the German *Encyklopädie der Mathematischen Wissenschaften*, as well as in the French edition, *Encyclopédie des Sciences mathématiques pures et appliquées* to the iteration of complex functions (see, respectively, Pincherle [1907] and [1912]).

At the time of the Academy's prize announcement in 1915, nothing new regarding the global iteration of complex functions had appeared since Fatou's 1906 note. However, the climate had become very hospitable to the development of a global theory of iteration. This was due primarily to two recent trends in French mathematics. The first of these was the recognition by many of the younger French mathematicians that set theory was a vital element of function theory, a development which had begun slowly with the importation of set theory into France in the 1890's. Set theory had by 1915 proven useful in a number of areas of mathematics, including the theory of complex iteration, as Fatou's note made abundantly clear. The second was Montel's theory of normal families, which he had successfully applied to the study of the convergence properties of series and sequences of complex functions, as well as to the body of work which grew out of the Picard Theorems.

Although Fatou and Julia were the first to apply the theory of normal families to the iteration of complex functions, it is entirely possible that members of the Academy's mathematical section felt that the theory of normal families might provide the means to advance the study of iteration. Indeed, it is not too difficult to see a connection between normal families and iteration, as is indicated in the following examples, which would have been familiar to anyone well-versed in the existing studies of iteration.

Example 9.1 *Let $\phi(z) = z^2$, and let*

$$\mathcal{G} = \{\phi^n : n \in \mathbb{Z}_0^+\}$$

where $\mathbb{Z}_0^+$ denotes the non-negative integers. Also let D be the open unit disc, J the unit circle and E the region $|z| > 1$.

The attracting fixed points of $\phi(z)$ are 0 and ∞. Since $\phi^n(z) = z^{2^n}$, it is evident that points in the open unit disc D converge to 0 under iteration by $\phi(z)$, while points in E converge to ∞. All points z on the unit circle J satisfy $|\phi^n(z)| = 1$, hence they remain in J under iteration by $\phi(z)$ and therefore do not converge to either of the attracting fixed points.

Let S be any subsequence from the iterative family $\mathcal{G}$ and let z be any point in $\bar{\mathbb{C}}$ not on J. Since all points not on J converge to either 0 or ∞, S converges uniformly to either the constant function $G(z) \equiv 0$ or to $G(z) \equiv \infty$, depending on whether z is in D or in E. Thus the family $\mathcal{G}$ is normal in some neighborhood of all points not in the unit circle.

However, if N is any neighborhood of a point z on the unit circle J, then the family $\mathcal{G}$ is not normal on N, since a subsequence S of functions from $\mathcal{G}$ does not converge uniformly on N to a meromorphic function. This can be seen directly as follows. Let S be a subsequence of $\mathcal{G}$ restricted to N. Since $\mathcal{G}$ is normal on D, the subsequence S converges uniformly to 0 on $N \cap D$ which implies that if S converges uniformly to a meromorphic function, this function must be $G(z) \equiv 0$. On the other hand, $\mathcal{G}$ is normal on E and a similar argument implies that the sequence S converges uniformly on $N \cap E$ to the meromorphic function $G(z) \equiv \infty$. Thus $\mathcal{G}$ is not normal on any neighborhood of points from J. In other words, the family $\mathcal{G}$ is normal on neighborhoods of points which converge to the attracting fixed points 0 and ∞, and non-normal on neighborhoods of points which do not converge to either of the attracting fixed points.

The Newton's method function $N(z) = (z^2+1)/2z$ for the quadratic $q(z) = z^2 - 1$, which was studied by Schröder in his papers [1870] and [1871] and Cayley in his paper [1879a] (see Chapter 1) has identical dynamics to the map $\phi(z) = z^2$ since it is analytically conjugate to $\phi(z)$. The fixed points 0 and ∞ of $\phi(z)$ correspond to the fixed points ± 1 of $N(z)$, and the sets D, J and E correspond respectively to the left half-plane, the imaginary axis and the right half-plane.

Example 9.2 *Let $\phi(z) = z^2/(z^2+2)$. Let $\mathcal{G}$ be as in the previous example.*

As was shown in the chapter concerning Fatou's 1906 note, $\phi(z)$ converges to the attracting fixed point 0 on the entire Riemann sphere except on a totally disconnected set J. Let A_0 be the set $\bar{\mathbb{C}} - J$. On a sufficiently small neighborhood of any point z in A_0, all subsequences from $\mathcal{G}$ converge uniformly to the constant function $G(z) \equiv 0$. On a neighborhood N of points z from J the same behavior seen in the above example is evident: no subsequence S from $\mathcal{G}$ can converge on N to a meromorphic function $G(z)$ because on discs contained in $N - J$ the sequence S converges to $G(z) \equiv 0$. However, since points in J do not converge to 0 under iteration by $\phi(z)$, the sequence S does not even converge pointwise to $G(z) \equiv 0$ on N. Thus S converges uniformly to no meromorphic function on N, and $\mathcal{G}$ is therefore not normal on neighborhoods of points in J.

These examples suggest a dichotomy that anyone interested in the study of iteration might think worth exploring: the iterative family $\mathcal{G}$ is normal on a neighborhood of any point which converges to an attracting fixed point under iteration by $\phi(z)$, but is not normal on any neighborhood of a point from the set of non-convergence

J.

At least two members of the Academy, Appell and Picard, had enough familiarity with both the study of complex iteration and Montel's work to be aware of the potential application of the theory of normal families to the theory of iteration. Montel's work in normal families had earned him the French Academy's *Prix Gustave Roux* in 1913, a prize whose purpose it was to reward young French mathematicians of exceptional promise. As members of the mathematics section of the Academy, both Appell and Picard would have had a hand in the awarding of this prize. Moreover, both Appell and Picard were also familiar with the study of iteration since they served with Koenigs on the thesis examination committees for both Auguste Grévy and Leopold Leau.

Given both the existence of a French body of work which concerned the iteration of complex functions and the familiarity on the part of members of the Academy with recent developments which might prove useful in the study of iteration, Poincaré's work in celestial mechanics would by no means be the only motivation for the Academy's decision to solicit papers on iteration. Further indication of this is the Academy's explicit suggestion that entrants confine themselves to the study of rational substitutions, which had been linked to the study of complex iteration in Fatou's note [1906a]. Finally, it is perhaps significant that, as was the case with Koenigs, neither Julia or Fatou attempted to link their studies to Poincaré's return map, despite the fact that both referred to Poincaré's work in areas not directly related to iteration, such as automorphic functions. Fatou and Julia, in contrast, each made frequent reference to their predecessors in the study of iteration, especially to the work of Koenigs.

Julia, in fact, never mentioned Poincaré's work on iteration in his paper [1918]. Fatou's only reference to it was to include Poincaré's paper [1890] in a listing of previous works which utilized iteration in one form or another. This paper did not concern Poincaré's study of mechanics, nor did it have much to do with iteration. It did, however, treat a functional equation, often called the Poincaré functional equation, which can be used in the study of repelling periodic points. Poincaré, however, did not use it in this fashion. Both Poincaré's paper and the Poincaré functional equation are discussed briefly in Chapter 10.

9.4 The Awarding of the Prize

The Academy announced at its December 2, 1918 meeting that Julia's paper [1918] was the winning paper. It announced also that two other mathematicians entered papers, one of whom, Lattès, was awarded honorable mention posthumously. Curiously, the Academy did not name the third entrant, and its sole reference to this entry was to note that it had retained only the manuscripts of Julia and Lattès.

The third entrant was the Italian mathematician Salvatore Pincherle, who, as was noted earlier in this chapter, wrote on functional equations in the *Encyklopädie der Mathematischen Wissenschaften.* His entry to the contest concerned the iteration of the function $p(x) = x^2 - a$, and based on several notes that he published on the subject around the same time, including [1917], [1918a] and [1918b], he focused on the case where a is positive, although he considered the case where a is complex in [1920a], and expanded his interests to include polynomials as well. His interest in this particular equation was motivated by his interest in investigating the expression

$$\pm\sqrt{a \pm \sqrt{a \pm \cdots \pm \sqrt{a}}} = x,$$

where a may appear infinitely many times. If, for example, a appears n times, x would be a root of $p^n(x)$. If the number of a's were infinite, and the expression were to converge, it would converge towards a fixed point of $p(x)$.

In awarding Julia the prize, the prize commission of the Academy noted the following:

> Julia's memoir bears the mark of a mathematical spirit of the highest order, which with vigor understands problems in the fullest generality and pursues the consequences as far as necessary; it shows equally a profound knowledge of the results and methods of modern analysis with a remarkable aptitude to utilize them. ...
>
> Thus, the commission unanimously advised that the *Grand Prix des Sciences mathématiques* be awarded to Gaston Julia; it proposes as well to give the work of Lattès, a most honorable mention.
>
> The Academy has adopted the propositions of the commission [*Comptes rendus*, December 2, 1918:814].[1]

In a separate notice from the same meeting, the Academy announced that Lattès had been awarded 2000 francs for his work in analysis and that Fatou had also been awarded 2000 francs for his contributions in both the theory of series and the iteration of rational functions. Perhaps not coincidentally, the Academy at this time also presented Montel with a prize of 1000 francs for his work concerning sets of analytic functions.

The prize announcement also contained a brief but interesting comparison of Fatou's work to that of Julia:

> At various points during his research, Julia had consigned his results in sealed letters, deposited at the Academy; after the deposit of these letters, on December 17, a well-known geometer, Fatou, who had already made interesting progress on his own in the study of iteration, announced a large portion of these results in *Comptes rendus*, which he obtained on his own but in the same manner [as Julia], in utilizing the properties of the normal sets of Montel: this is not the first time in the history of

[1] Members of the commission included Appell, Picard, Jacques Hadamard and Humbert.

> the Science that two worthy investigators arrived at the same time, by the same march, at the same discoveries [*Comptes rendus*, December 2, 1918:814].

As the quotation directly above indicates, on December 17, 1917 Fatou announced several results which he had discovered in the course of his investigations into the iteration of rational functions. Matters became complicated two weeks later when Julia announced that prior to the appearance of Fatou's note, he had established identical results, and offered proof of the matter in the form of the following letter, which appeared in the *Compte rendu* of December 31, 1917:

> I read with interest the note of Fatou published in the *Compte rendu* of December 17, 1917. The essential results which it contained, I had previously entered in a series of four sealed letters which I have deposited with the Secretary of the [Academy] and which were registered with numbers 8401, 8431, 8438, 8466, dated, respectively, June 4, 1917, August 17, 1917, September 17, 1917 [and] December 10, 1917. The Academy can judge, upon opening these letters, whether they contain the results given by Fatou ... The Academy can assess, at the same time, whose methods and whose results ought to be given priority [Julia 1917: 105].

The Academy heeded Julia's request that it open these letters and judge priority, for immediately following Julia's letter was a response entitled "On a Communication of Gaston Julia" written by Humbert. He reported that in response to "questions raised [by Julia] concerning a question of priority," he had opened Julia's sealed letters, and that "the assertions of Julia are founded" since the letters do indeed contain "all the results for which he has claimed priority [Humbert 1917:107]."

In the folk history of mathematics this incident has grown to epic proportions, and it is sometimes said that Julia publicly accused Fatou of stealing his results. Although Julia made no such claims in his letter, he referred to the similarity of Fatou's approach as a "curious coincidence." And although it would be an exaggeration to term this note a personal attack on Fatou, Julia did take a rather superior tone, and observed, with some justification, that his approach was both more precise and more general.

Fatou's longstanding interest in the iteration of rational functions, and the fact that at the time he was at work on an extensive treatise on the subject, suggest that he had intended to enter the contest. Perhaps Humbert's comments regarding Julia's priority dissuaded him. I do not mean to suggest that the Academy displayed any favoritism towards Julia—after all, it is important to keep in mind a characterization of Julia's work already quoted from the Academy's announcement: "Julia's memoir is marked by a mathematical spirit of the highest order." Nonetheless, it is interesting to speculate whether patriotic sentiments or personal contacts would have had any influence on the prize deliberations had Fatou decided to enter his paper. As the reader will recall Julia was a war hero, and was visited from time to time by both Picard and Humbert during his recuperation from his battle wounds.

Before turning to the work of Fatou and Julia, I will in the next chapter briefly discuss Lattès' approach, as well as that of the American mathematician Joseph Fels Ritt (1893–1951), who wrote several papers on iteration around 1918, and whose view of iteration was quite similar to that held by Lattès. Although the studies of these mathematicians represent an alternative to the approach of Fatou and Julia, it is in one sense a rather impoverished one, since neither came close to duplicating the elegance or the accomplishments of Fatou and Julia.

Chapter 10

Lattès and Ritt

10.1 Biographical Sketch of Ritt

Joseph Fels Ritt was born six months after Julia in 1893. He received his doctorate from Columbia University in 1917 for his work on differential operators, written under the supervision of Edward Kasner (1878–1955). Following World War I, Ritt taught at Columbia until his death in 1951. Although he wrote several articles on iteration in the late teens and early twenties, his chief interests involved the study of differential equations, and both he and his students made many important contributions to the field of algebraic differential equations.

The results summarized in Ritt's note [1918] are rather modest, as is the paper [1920] in which he detailed the results presented in this note. Together they illuminate a small portion of the problem posed by the Academy, namely, the behavior of iterates in the neighborhood of a repelling fixed point.

Ritt's work is of interest for two reasons. His approach is virtually identical to the one Lattès took in studying the iteration of complex functions of a single variable, and together their work gives us a glimpse of what the study of complex dynamics might look like without the application of Montel's theory of normal families. Secondly, Ritt's note is indicative of an emerging American interest in iteration. In the second decade of the twentieth century, several Americans, including Ritt's thesis advisor Kasner, touched upon several issues relating to iteration and functional equations. Some of this work appeared prior to the Academy's announcement, so the contest alone was not responsible for Ritt's interest. However, as was the case with Ritt's work, their approach to the subject had much more in common with the Koenigs school than with the work of Fatou and Julia.

10.2 The Approach of Lattès and Ritt

For both Ritt and Lattès the study of the iteration of complex functions of one complex variable revolved around the functional equation

$$F(su) = \phi(F(u)), \tag{10.1}$$

where u is a complex variable, s is a complex constant satisfying $|s| > 1$, and $\phi(z)$ is a given rational function. They found this equation particularly useful to study iteration on a neighborhood of a repelling fixed point of $\phi(z)$, that is a fixed point whose multiplier s is strictly greater than one in modulus.

This equation is a special case of the following system of functional equations, discussed by Poincaré in his paper [1890]:

$$\begin{aligned} F_1(su) &= \phi_1[F_1(u), \ldots, F_n(u)] \\ \vdots &= \qquad \ddots \\ F_n(su) &= \phi_n[F_1(u), \ldots, F_n(u)]. \end{aligned}$$

The $\phi_i(u_1, \ldots . u_n)$ are rational functions from $\bar{\mathbb{C}}^n$ to $\bar{\mathbb{C}}$, and the $F_i(u)$ are entire or meromorphic functions on $\bar{\mathbb{C}}$. Poincaré's interest in this system of equations was motivated by his study of Cremona substitutions, which are bi-rational transformations on $\bar{\mathbb{C}}^n$.

Because equation (10.1) is a special case of the system of equations Poincaré introduced, it is generally referred to as the Poincaré functional equation, even though Poincaré did not use it to study iteration, and Ritt and Lattès used it in a manner he evidently did not foresee.

Ritt and Lattès found the Poincaré equation a useful tool to study of iteration for several reasons. For example, if $\phi(x) = x$ and $s = \phi'(x)$, with $|s| > 1$, then a solution $F(u)$ to (10.1) exists in a neighborhood of x, and the function $F(u)$ can be extended to a meromorphic function on the complex plane. Therefore, by Picard's Little Theorem, $F(u)$ has at most two exceptional values, that is, there are at most two points z_0 and z_1 such that there is no u which satisfies either $F(u) = z_0$ or $F(u) = z_1$. If exceptional points exist, they are attracting periodic points of $\phi(z)$. The fact that $F(u)$ is defined globally also makes the Poincaré equation a potentially handy means to study the iteration of arbitrary points in the plane. As long as z is not an exceptional value of $F(u)$, there exists a u such that $F(u) = z$. Since

$$\phi(z) = \phi(F(u)) = F(su),$$

it follows that

$$\phi^2(z) = \phi(F(su)) = F(s^2u),$$

and consequently,

$$\phi^n(z) = F(s^n u). \tag{10.2}$$

Thus the Poincaré equation reduces iteration by $\phi(z)$ to repeated multiplication of u by s.

The reader may recall that the canonical Schröder equation

$$B(\phi(z)) = \phi'(x)B(z), \tag{10.3}$$

considered in the neighborhood of a fixed point x of $\phi(z)$, also reduces iteration to repeated multiplication by a constant, but it should be kept in mind that, at the time of Lattès' writing, $B(z)$ was generally defined only on a neighborhood of a fixed point.

Lattès noted as well in [1918a] that the Poincaré functional equation is the inverse of the Schröder functional equation in the sense that, if $s = \phi'(x)$ and $|s| > 1$, then the solution $F(u)$ to equation (10.1) is the inverse of the solution $B(z)$ to the Schröder equation. Lattès also pointed out that $B(z)$ is generally one element of a multi-valued function, while $F(z)$ is a single-valued, meromorphic function, and therefore $B(z)$ generally can not be extended to a globally defined, single-valued function. Thus the solution $B(z)$ to the Schröder equation is the multi-valued inverse of $F(u)$. That $B(z)$ is multi-valued when $|\phi'(x)| > 1$ follows from Koenigs' solution of the Schröder equation discussed in Section 3.6, which he solved by inverting the solution to the canonical Schröder equation for $\psi(z)$, where $\psi(z)$ is an inverse of $\phi(z)$ satisfying $\psi(x) = x$.

As the following quotation indicates, Lattès believed that $F(u)$, being a globally defined function, gave the Poincaré equation a clear edge over the Schröder equation since it could be used to study both local and global questions:

> The advantage of substituting the Poincaré equation for the Schröder equation is that [except for the exceptional values of $F(u)$] one can study the iteration of any initial value z in the plane. ...
>
> *The problem relating to iteration of a given substitution is thus transformed to the problem relative to the growth [i.e., the iteration] of $F(u)$.* ... This problem, in its general form, can be stated thusly:
>
> *Determine the [derived] set E' of the set E of the consequents z_n of a point z arbitrarily given* [1918a:27].

The italics are Lattès'.

An interesting example which neither man gave but which sheds some light on the approach favored by Ritt and Lattès involves Schröder's proof of the convergence of Newton's method for the quadratic, discussed in Section 1.7. As the reader will recall, Schröder demonstrated the convergence of the Newton's method function $N(z)$ for the quadratic $q(z) = z^2 - 1$ to the roots of ± 1 of $q(z)$ by demonstrating that the function $M(z) = 2z/(z^2 + 1)$, which is conjugate to $N(z)$, converges to -1 under the iteration of arbitrary points in the left half-plane and to 1 for points in the right half-plane.

Schröder accomplished this by exploiting the identity

$$M(z) = \frac{2z}{z^2+1} = -i\tan(2\arctan(iz)), \tag{10.4}$$

from which it follows that

$$M^n(z) = -i\tan(2^n\arctan(iz)). \tag{10.5}$$

He also noted that (10.4) is equivalent to the functional equation

$$B(M(z)) = 2B(z), \tag{10.6}$$

where, fixing a branch of the arctangent, $B(z) = \arctan(iz)$.

As this book has progressed, the reader has perhaps noticed that functional equation (10.6) is not at all like the canonical Schröder equation (10.3) used to investigate iteration near an attracting fixed point x. Unlike the usual local solution to the canonical equation, Schröder's solution to (10.6) was not generated on a neighborhood of an attracting fixed point of $M(z)$ but was rather defined globally, enabling him to evaluate (10.5) for arbitrary points on the plane. Moreover, the constant 2, having a modulus greater than 1, is not the multiplier of an attracting fixed point.

A look at a particular Poincaré equation for $M(z)$, which can be easily generated from equation (10.6), reveals what is really going on with Schröder's equation for $M(z)$, at least from the vantage point Ritt and Lattès developed. Replacing z by $F(u) = -i\tan(u)$ in equation (10.6) yields

$$B(M(F(u))) = 2B(F(u)),$$

which reduces to

$$B(M(F(u))) = 2u$$

since $B(F(u)) = \arctan(i(-i\tan(u))) = u$. Applying $F(u)$ to both sides of the above equation, and rearranging terms, we get

$$F(2u) = F(B(M(F(u)))),$$

which for similar reasons reduces to

$$F(2u) = M(F(u)), \tag{10.7}$$

the Poincaré equation for $M(z)$. Moreover, it follows from a simple computation that 0 is a repelling fixed point of $M(z)$ satisfying $M'(0) = 2$, so equation (10.7) is the canonical Poincaré equation generated around a repelling fixed point of $M(z)$, and the functional equation Schröder considered is therefore equivalent to the Poincaré equation. Consequently, from the point of view of Lattès and Ritt what Schröder actually did was to solve the Poincaré equation around a repelling fixed point of

$M(z)$, and because such solutions are globally defined, was able to show that $M(z)$ converges to either ± 1 by directly evaluating $F(2^u) = \tan(2^n u)$.

This utilization of a functional equation to study the global properties of iteration was precisely the sort of approach Lattès advocated in the above quotation. The irony of it all is that in showing the convergence of $M(z)$ to its attracting fixed points, which Schröder felt was a vindication of his theory of attracting fixed points, he was, in fact, taking advantage of what Lattès and Ritt later revealed to be a property of repelling fixed points.

However, there is a problem with basing the study of iteration on the derived set E' of the set

$$E = \{\phi^n(z_0)\} = \{F(s^n u_0)\}$$

as Lattès suggested in the above quotation, and it is a problem which Schröder faced as well, namely, that the set E' can, depending on the nature of the function $F(u)$, be quite difficult to determine. It should come as no surprise, then, that neither Lattès nor Ritt developed a satisfactory method of determining the set E' for arbitrary functions. In fact, the only general result they proved regarding the set E' is that, being closed, E' can be decomposed into a perfect set and a denumerable set, either of which may be empty.

Lattès, however, indicated he was aware of the difficulties involved in determining E' and suggested in [1918a] that one approach to the Poincaré equation is to choose as $F(u)$ functions for which $F(mu)$, for integer m, is a rational function of $F(u)$. In this manner he hoped to find a rational function $\phi(z)$ which satisfies the Poincaré functional equation $F(mu) = \phi(F(u))$ for that particular $F(u)$. For example, he noted in [1918a] that if $F(u) = \tan(u)$, the fact that

$$\tan(2u) = \frac{2\tan(u)}{1 - \tan^2(u)},$$

implies that

$$F(2u) = \phi(F(u)),$$

where $\phi(z) = 2z/(1 - z^2)$. From here, the form of $\phi^n(z)$ would presumably be deduced via evaluation of $F(2^n u) = \tan(2^n u)$.

If this approach seems vaguely familiar it is because Schröder used a similar means in discovering the relationship at (10.4). It also echoes Schröder's suggestion, discussed in Section 1.6, to turn things around and rather than solve the Schröder equation $B(\phi(z)) = sB(z)$ directly for $B(z)$, to instead treat $B(z)$ as known and $\phi(z)$ as unknown, and search for functions $\phi(z)$ satisfying $B(\phi(z)) = sB(z)$ via consideration of the family

$$B^{-1}(sB(z)),$$

as $B(z)$ varies.

These parallels with the work of Schröder serve to emphasize the fact that the approaches of Lattès and Ritt did not represent a truly fresh approach but instead were a continuation of the nineteenth century tradition of studying iteration by seeking the right functional equation. What Lattès and Ritt in essence did was to substitute for the Schröder equation what they felt was a more suitable functional equation. Without doubt their work served as an interesting coda to the theory set forth by the Koenigs school, and had the studies of Fatou and Julia not appeared, their contributions to the study of the iteration of rational functions of a single variable would have been quite valuable.

However, Ritt and Lattès were unable to answer the basic questions which vexed Koenigs and his successors: How is the plane partitioned into regions of attraction? What does the boundary of adjacent regions of attraction look like? The failure of Ritt and Lattès to answer these questions is not surprising in light of the fact that they did not take advantage of Paul Montel's theory of normal families, which served as the cornerstone of the studies of Fatou and Julia.

It is actually a little curious that neither Ritt nor Lattès thought to examine iteration from the point of view of normal families since one of their discoveries readily suggests it. Both men recognized that the function $F(u)$ satisfying the Poincaré equation is either an entire or meromorphic function, and explicitly noted that Picard's Little Theorem implies that $F(u)$ has at most two exceptional values. This in turn implies that for a given z, the set of iterates of $\phi(z)$, namely, the set $\{\phi^n(z)\}$, which both men studied, also has at most two exceptional values. It therefore follows immediately from Montel's normality criterion, proved in 1912 and stated above as Theorem 8.8, that the set of iterates, viewed as a family of functions defined on neighborhoods of repelling fixed points, is not a normal family.

If one were familiar with Montel's theory, which by 1918 had been proven useful in a wide variety of situations within complex function theory, particularly in connection with the exceptional values given by Picard's Theorems, it would not have been an extraordinary leap to connect the exceptional values of $F(u)$ to the theory of normal families. One wonders if perhaps either Fatou or Julia first investigated iteration from the point of view of the Poincaré functional equation and, in the manner suggested above, thought to apply Montel's theory.

It seems, however, that neither Ritt nor Lattès made this connection, at least not prior to the appearance of normal families in the work of Fatou and Julia. Lattès briefly mentioned the theory of normal families in his third note, [1918c], but it was in reaction to the presence of this theory in the publications of Fatou and Julia.

That Ritt did not use the theory of normal families is in keeping with his mathematical tastes, which ran to the classic. According to Lorch [1951], Ritt's mathematical style was rather conservative, and despite his considerable accomplishments he was surprisingly resistant to recent advances in mathematics. Indeed, it appears that Ritt might have had much more in common with the Hermitian old guard than with Fatou, Julia or Montel, despite the fact that he was younger than any of them.

Lorch notes that Ritt "begrudgingly recognized" the Lebesgue integral "but that fields of characteristic p were special objects of scorn [1951:310]." Ritt was evidently quite bothered that such fields were crucial to the study of his favorite subjects, number theory.

Chapter 11

Fatou and Julia

11.1 Introduction to the Studies of Fatou and Julia

Although the mathematical content of the respective approaches of Fatou and Julia to the iteration of rational functions is quite similar, there are considerable differences in both style and emphasis. Julia on the whole argued more precisely than Fatou. He presented his results in a more organized fashion, and he did a better job of utilizing important theorems from the theory of complex functions. Their works differ on another more subtle level which has as much to do with aesthetics as with mathematics. Fatou wrote in a gently meandering style, reminiscent of a certain nineteenth century style of mathematics, while Julia's paper is closer to the axiomatic style which predominates in contemporary mathematics.

Fatou's paper also recalls the nineteenth century in its emphasis on functional equations. While Fatou's major concern was certainly iteration, he by no means neglected the study of functional equations in his three-part work [1919], [1920a] and [1920b], as its title *Sur les équations fonctionnelles* indicates. Fatou used arguments involving functional equations throughout his work, and the last chapter of the final installment, [1920b], is devoted entirely to the study of functional equations.

Although Julia studied functional equations extensively in subsequent papers, he rarely used them in [1918] and clearly subordinated the study of functional equations to that of iteration, as the following quotation from his introduction to [1918] indicates:

> There remain, as has already been indicated, many other questions to treat. I could apply the results obtained herein to the well-known functional equations from the work of Koenigs and his successors. That study does not fall within the purview of this memoir, which is the study of iteration in and of itself: I may well ultimately return to questions

thus left suspended [1918:128].

I do not mean to leave the impression that Fatou's work is not first rate, for it is. Not only is his study of iteration remarkably similar in content to Julia's, but in some instances, he anticipated contemporary interests in ways Julia did not, such as his use of symbolic dynamics to investigate totally disconnected perfect sets [1919:257ff] and his tantalizing initial exploration of parameter space [1919:258ff].

11.2 Iteration and the Theory of Normal Families

Despite the differences in approach and style, the mathematical content of Julia [1918] and Fatou [1919], [1920a] and [1920b] is remarkably similar. Much of this similarity stems from the fact that each based his approach on Montel's theory of normal families.

In Examples 9.1 and 9.2, Montel's theory of normal families gives a very useful way to characterize the behavior of the iterates of a given function $\phi(z)$ for arbitrary points in the extended plane $\bar{\mathbb{C}}$. In these examples $\bar{\mathbb{C}}$ was partitioned into sets on which points either converge to an attracting orbit or converge to no attracting orbit. For example, if for all $z_0 \in D$, the sequence $\{\phi^n(z_0)\}$ converges to a fixed point x, then the family

$$\{\phi^n : n \in \mathbb{Z}_0^+\}, \tag{11.1}$$

where $\mathbb{Z}_0^+$ denotes the set of non-negative integers, was shown to be normal on D, and in fact converges uniformly to the constant function $G(z) \equiv x$. If, on the other hand, z is from the set of points J which converge to no attracting orbit, then the family (11.1) is normal on no neighborhood N of z because the behavior of the sequence $\{\phi^n(z_0)\}$, where $z_0 \in N$, varies dramatically with the choice of z_0.

These examples suggest that for an arbitrary function $\phi(z)$ it might be useful to partition $\bar{\mathbb{C}}$ into regions where family (11.1) is normal and non-normal. Indeed, this dichotomy is at the core of both Fatou's and Julia's studies. Both mathematicians divided $\bar{\mathbb{C}}$ into regions of normality and non-normality, which are often called, respectively, the Fatou and Julia sets. These sets are defined as follows:

Definition 11.1 *Let $\phi(z)$ be a rational function, and let $\mathcal{G}$ be the family*

$$\mathcal{G} = \{\phi^n : n \in \mathbb{Z}_0^+\}. \tag{11.2}$$

The Julia set *J is the set of points in $\bar{\mathbb{C}}$ for which there exists no neighborhood on which the family $\mathcal{G}$ is normal. The* Fatou set *F is the set of points for which there exists neighborhoods on which $\mathcal{G}$ is normal.*[1] *The set F is therefore the complement*

[1] I've taken the liberty of standardizing Fatou's and Julia's notation for the Julia set. Julia denoted the Julia set E' and Fatou referred to it as $\mathcal{F}$. I've also standardized their function notation.

of J in $\bar{\mathbb{C}}$. The sets J and F are also referred to as the domains of non-normality and normality, respectively.

It also can be shown that both sets are completely invariant, i.e., forward and backward invariant, under iteration by $\phi(z)$.

Generally speaking, points in F converge under iteration to attracting orbits, while points in J do not.[2] Instead, J is generally the boundary of the various regions of the plane which are attracted to periodic orbits, although there are functions, to be discussed below, for which J equals $\bar{\mathbb{C}}$.

Perhaps because they felt that by understanding the Julia set J, which forms the boundary of F, they would gain insight into the structure of F, both Fatou and Julia began their investigation of the dichotomy between the regions of normality and non-normality with a thorough investigation of the set J.

In what follows, the function $\phi(z)$ will be assumed to be a rational function of degree strictly greater than one.[3] Fatou and Julia each made this restriction, not only because the behavior of a linear fractional transformation (LFT) under iteration was understood, but because many of the following theorems do not hold for an LFT. For example, the Julia set of an LFT generally consists of a single repelling fixed point, while this is never the case for a rational function of degree greater than one.

11.3 The Julia Set

The following theorem, proved by both Fatou [1920a:33ff] and Julia [1918:157ff], gives a precise description of the Julia set.

Theorem 11.2 *Let $\phi(z)$ be a rational function. The Julia set of $\phi(z)$ is the closure of*

$$T = \left\{ x : \phi^p(x) = x \text{ and } \left| \frac{d\phi^p}{dz}(x) \right| > 1 \right\}, \tag{11.3}$$

for $p = 1, 2, \ldots$. Moreover, the Julia set is perfect.[4]

[2] For certain rational functions which possess a fixed point x whose multiplier is one in modulus but not a root of unity, there are components of F which do not converge under iteration to a period p orbit. Fatou called such domains singular domains, and more will be said about them below. However, points in J never converge to an attracting periodic orbit.

[3] The degree of a rational function $\phi(z) = p(z)/q(z)$ is the maximum of the numbers degree $p(z)$ and degree $q(z)$.

[4] Julia actually started his investigation of the domain of non-normality by defining T and then proving that its closure equals J. Fatou, on the other hand, began his formal investigation of the Julia set, as I have done, by first partitioning $\bar{\mathbb{C}}$ into domains of normality and non-normality, and then proving Theorem 11.2.

Since period p points x for which

$$|\frac{d\phi^p}{dz}(x)| > 1$$

are called *repelling periodic points*, this theorem asserts that the Julia set is the closure of the repelling periodic points.

Julia proved that the iterative family $\mathcal{G}$, defined at (11.2), is not normal on J by first showing that $\mathcal{G}$ is neither normal on T nor on its closure.[5] He then showed that the set J on which $\mathcal{G}$ is not-normal is precisely the closure of the set T. I will outline the portion of the proof in which Julia showed by contradiction that the set J contains the closure of the set T (see [1918:163ff]).

First let $x \in T$ satisfy $\phi(x) = x$ with $|\phi'(x)| > 1$. Suppose as well that N is a neighborhood of x on which $\mathcal{G}$ is normal. Without loss of generality assume that $x \neq \infty$ and that $\phi(z)$ has no poles on N. (If this is not the case, use the coordinate change $z \mapsto 1/z$.) Let $\{\phi^{n_i}\}$ be a subsequence of $\mathcal{G}$ which converges uniformly on N to a limit function $G(z)$. The function $G(z)$ is either analytic or uniformly infinite on N. Since x is a finite fixed point of $\mathcal{G}$, $G(x) = x \neq \infty$, and $G(z)$ is therefore analytic. However, $G'(x) = \infty$, because $|\phi'(x)| > 1$, and, as can be shown via a direct calculation involving the chain rule,

$$\left|\frac{d\phi^{n_i}}{dz}(x)\right| = |\phi'(x)|^{n_i},$$

so $|\phi'(x)|^{n_i}$ goes to ∞ with n_i. This contradicts the analyticity of the limit function $G(z)$, hence the family $\mathcal{G}$ is not normal around points in T.

The extension of the argument to the case where x is a period p point in T follows directly by applying the $p = 1$ case to the function $\phi^p(z)$. Moreover, since $\mathcal{G}$ is not normal on T, $\mathcal{G}$ is not normal on the closure of T because any neighborhood N of points in the closure of T contains points from T. Blending topological notions with a normal families argument, Julia next showed that the closure of J is perfect, and concluded his proof of Theorem 11.2 with the demonstration that if $\mathcal{G}$ is not normal around a point, then that point is in the closure of T.

In light of the connection between Picard theory and Montel's theory of normal families which was discussed in a previous chapter, it is worth noting that Julia followed his first proof that J is a perfect set with a second proof utilizing a version of Schottky's Theorem, presented earlier as Theorem 8.9, which he called the Picard-Landau Theorem, and afterwards remarked

> This second demonstration is not in essence different from my first, especially when one realizes that Montel deduced the theorem of Picard-Landau from notions he introduced regarding normal families of functions [1918:171].

[5] For the sake of brevity, I will often use the phrase "normal on X" when I really mean "normal on neighborhoods of X" or "normal on neighborhoods of points in X."

11.4 Some Interesting Julia Sets

Fatou and Julia were both intrigued by the fact that for certain functions, the Julia set could be quite unusual. For example, as Fatou indicated in his note [1906a], the Julia set could be a totally disconnected perfect (TDP) set. Both Fatou and Julia showed that the set J could also be a continuous curve $\gamma(t)$ for which $\gamma'(t)$ does not exist at an infinite number of points ([Julia 1918:273ff], [Fatou 1920a:91]), a continuous curve with an infinite number of double points, that is, a curve which crosses itself infinitely often ([Julia 1918:232ff], [Fatou 1920a:89]), a closed Jordan curve ([Julia 1918:212ff], [Fatou 1920a:260ff]), a segment ([Julia 1918:260]), or a set consisting of infinitely many disjoint continuous pieces ([Julia 1918:257ff], [Fatou 1920a:87]).

Examples of TDP sets, curves without tangents, and curves which have infinitely many double points existed in the mathematical literature of the time, but were generally given by a detailed and artificial constructive process. Moreover, as was noted in the chapter concerning Fatou's 1906 note, many French mathematicians were disturbed by the existence of such things as curves without tangents and not only regarded them as unnatural but sometimes ridiculed those who studied them. Perhaps as a rejoinder to such sentiments, Fatou and Julia each provided several constructions of these sorts of sets and curves, and the fact that they occurred so frequently as boundaries of the Fatou set offered persuasive evidence that such things were by no means unnatural.

Julia explicitly noted that the fact that J can be viewed simply as the domain of non-normality or the closure of the repelling period p points gave him a simple means of constructing interesting sets. In comparing Julia sets with infinitely many undefined tangents with curves produced by Helge von Koch (1870–1924), Fatou remarked that

> These curves have many similarities with those of Helge von Koch; but the curves he studied are defined in a constructive manner in which properties are assigned in advance [Fatou 1920b:242].

Among the curves to which Fatou refers is the so-called Koch Snowflake which Koch presented in his paper [1906] evidently as an example of a non-differentiable, non-rectifiable continuous curve which encloses a finite area. It is formed recursively from an equilateral triangle Δ_0 of side s by first deleting the middle third of each side and then replacing it with an equilateral triangle of side $s/3$ as indicated in figure 11.1. The resulting figure Δ_1 has twelve sides of length $s/3$. Next, in the same fashion, replace the middle third of each of the twelve sides with equilateral triangles of side $s/9$. The Koch Snowflake is the limiting curve obtained if the process is repeated indefinitely. In another example Julia noted that ideas which "arise naturally" in the study of iteration offer the "*simplest possible*" [Julia's emphasis] illustrations of

> the most subtle notions concerning the frontiers of planar, simply con-

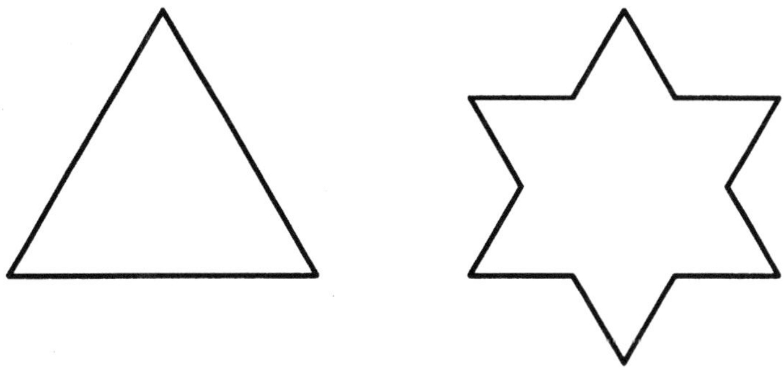

Figure 11.1: *Steps I and II in the construction of The Koch Snowflake.*

nected domains, ideas which have been recently discussed in the interesting memoirs [of Lindelöf and Montel] ... [1918:249].

What Julia evidently refers to in these remarks is the fact that for the function $\phi(z) = (3z - z^3)/2$, the Julia set is a simply connected curve on the Riemann sphere. This provides a natural example of a simply connected domain whose boundary is a continuous curve with an infinite number of double points, ideas which he pointed out are discussed in Lindelöf [1915] and Montel [1917]. As a schematic realization of the curve he offered a recursive construction which he noted was based on ideas from Koch [1906]. Julia's schematic is presented at figure 11.2.

Fatou also emphasized that the Julia set provides interesting examples of sets previously thought to be somewhat abnormal. Using a technique that both he and Julia sometimes employed, Fatou perturbed the coefficients of a given function whose Julia set was a TDP set, and observed that in the perturbed function, J remains a TDP set.[6] Julia likewise indicated that if J is a curve $\gamma(t)$ such that $\gamma'(t)$ does not exist at an infinite number of points, then this condition persists under perturbations of the coefficients of $\phi(z)$ [1918:292ff].

Fatou's argument that under certain perturbations J remains a TDP set is particularly interesting since it anticipates the contemporary study of the Mandelbrot set, which is the set of points c such that the Julia set of the function $q_c(z) = z^2 + c$ is connected. If c is not in the Mandelbrot set, the corresponding Julia set is a TDP set.

Using the function $\phi(z) = z^m + c$ as an example, Fatou considered the effect that varying the complex parameter c had on the structure of the set J. He noted initially

[6]Fatou claimed in his note [1906a] that the Julia set for certain functions did not have a well defined tangent at an infinite number of points. However, the justification he provided for this statement in [1906a] was rather vague. In [1920b] he proved that if the Julia set of a function $\phi(z)$ is a curve $\gamma(t)$ and if x is a period p point of $\phi(z)$ with $\gamma(t_0) = x$, then $\gamma(t_0)$ does not have a well-defined tangent whenever $\frac{d\phi^n}{dz}(x)$ is non-real.

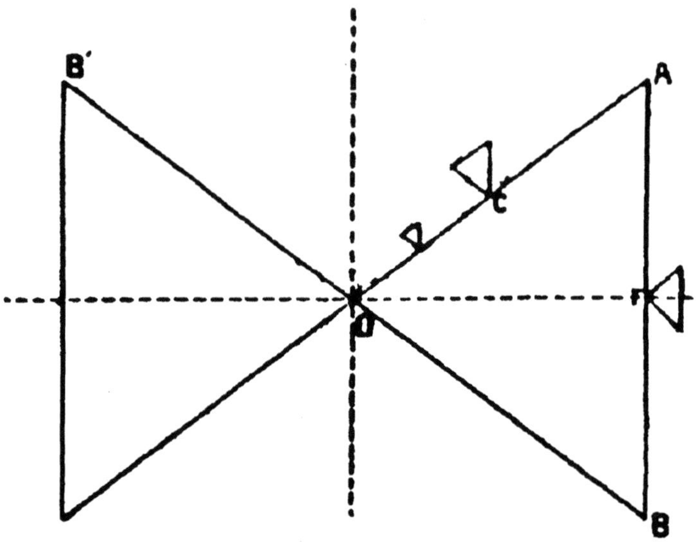

Figure 11.2: *Julia's schematic of the Julia set for the function* $\phi(z) = (3z - z^3)/2$ *[Julia 1918:244].*

that if $|c|$ was "sufficiently large," then J was a TDP set [Fatou 1919:254]. He later gave more precise estimates on $|c|$ in the more general case where $\phi(z) = zP(z) + c$ where $P(z)$ is a polynomial [1919:258–59]. Although he did not do so, his estimate applied to the special case of $q_c(z) = z^2 + c$ implies that if $|c| > 1 + \sqrt{6}/2$ then J is a TDP set. Compared with the standard contemporary estimate that $|c| > 2$, Fatou's estimate was fairly accurate.

That J remains a TDP set under such broad conditions on c indicated to Fatou that the "pathological" condition that J be a TDP set was actually quite normal, as the following quotation by Fatou suggests:

> Another remark that I'd like to make is that the existence of domains of convergence whose frontiers [i.e., J] are totally disconnected sets is not a singular case, that is, this condition can be produced without particular relationships among the coefficients of the function $\phi(z)$; it suffices that the coefficients vary in a convenient domain [1919:258].

11.5 Further Properties of the Julia Set

Recall from Definition 8.7 that if $\mathcal{F}$ is a family of meromorphic functions on a domain D, then a point w_0 is called an exceptional value of the family $\mathcal{F}$ on D if

$$\bigcup_{f\in\mathcal{F}} f[D]$$

does not contain w_0. Montel's normality criterion, Theorem 8.8, states that if a meromorphic family $\mathcal{F}$ has at least three exceptional values, then it is normal, hence a non-normal family of meromorphic functions can have at most two exceptional values. Julia [1918:167] and Fatou [1920a:35] used both Montel's normality criterion and the fact that $\mathcal{G}$ was not normal on neighborhoods of the Julia set to prove the following fact about J:

Theorem 11.3 *Let $\phi(z)$ be a rational function of degree strictly greater than one. With the exception of at most two points, given any point z in $\bar{\mathbb{C}}$, J is contained in the closure of the set $O^-(z)$, where $O^-(z)$ denotes the set of preimages of z under ϕ.*[7]

The proof I will offer is based on Fatou's proof. It uses the fact, which I will not prove, that because the family $\mathcal{G}$ consists of the iterates of $\phi(z)$, the set of exceptional values $\mathcal{E}$ of the family $\mathcal{G}$ for a given point $z \in J$ is the same for all sufficiently small neighborhoods of z. Since it can also be shown that the set of exceptional values is independent of the choice of $z \in J$, it is appropriate to refer to *the* exceptional set $\mathcal{E}$ of a given rational function $\phi(z)$.

With these facts in mind, let z be a point in J, and let N_z be a neighborhood of z which contains no exceptional points. As noted above, the family $\mathcal{G}$ is not normal on N_z. Let w be a point on the sphere which is not in $\mathcal{E}$. Then, since $\mathcal{G}$ is not normal on N_z, and w is not an exceptional point,

$$w \in \bigcup_{n=1}^{\infty} \phi^n[N_z].$$

Therefore, there is a point w' in N_z and an integer n such that $\phi^n(w') = w$. Since w' is by construction not an exceptional point, the argument applied to w can now be applied to w', hence there is a point w'' in N_z which is the preimage of w'. Continuing in this manner, N_z contains an infinite number of preimages of w under $\{\phi^n(z)\}$. Since this holds even if the neighborhood N_z is made arbitrarily small, z is in the closure of $O^-(w)$. This argument can be repeated for any point z in J, hence J is a subset of the closure of $O^-(w)$.

Theorem 11.3 was not only useful for Fatou and Julia, but has provided the means for contemporary researchers to construct computer representations of Julia

[7] The set $O^-(z)$ is often called the backward orbit of z under $\phi(z)$.

sets: taking an arbitrary point w in the plane, the Julia set is approximated by

$$\bigcup_{n=I}^{N} \psi^n(w),$$

where I and N are conveniently chosen, and where a different inverse $\psi(z)$ of $\phi(z)$ is chosen at random for each n. Choosing an appropriate I gets rid of what are called transients, that is, preimages of $\phi(w)$ which are relatively far from the Julia set.

Fatou and Julia were also intrigued by what a present day observer might call the self-similarity or fractal properties of the Julia set, which they studied via the following extension of Theorem 11.3:

Theorem 11.4 *Let $\phi(z)$ be a rational function of degree strictly greater than one. Let $z \in J$ and let N be a neighborhood of z which is bounded away from the set $\mathcal{E}$ of the exceptional values of $\mathcal{G}$. Let S be any closed set in $\bar{\mathbb{C}}$. There exists n and k such that*

$$N \subset \phi^n[N] \subset \phi^{2n}[N] \subset \cdots \subset \phi^{kn}[N] \cdots \qquad (11.4)$$

with $S \subset \phi^{kn}[N]$.

Fatou and Julia used Theorem 11.4 to demonstrate what Fatou called the homogeneity of J. The set A is said to be ***homogeneous*** if given any neighborhood N of a point in A there exists a continuous map $\Phi(z)$ such that $\Phi[N \cap A] = A$. That such a map exists for the set J follows directly from Theorem 11.4: choose z and N as in the statement of the theorem and let J be a closed set S. There then exists k and n such that $\phi^{kn}[N \cap J]$ contains J. The invariance of $\phi(z)$ on both F and J implies that $\phi(z) \in J$ if and only if $z \in J$, hence the last sentence of the theorem implies that $\phi^{kn}[N \cap J] = J$. The interest that homogeneous sets held for Fatou and Julia was expressed succinctly by Fatou: "The set J has the same structure in all of its parts [Fatou 1920a:40]." Julia noted that the homogeneous character of J implied that

> One can say that from any small portion of J, one can generate J in its entirety in a finite number of iterations [via the map $\phi^{kn}(z)$]. ... The structure of J in its entirety is the same as in any of its parts [Julia 1918:173n].

Especially interesting to these men was the fact that the global topological structure of J is mirrored in its local structure. Both mathematicians pointed out that the homogeneity of J implies that if $N \cap J$ is totally disconnected, then so is J. Likewise, if $N \cap J$ is continuous, J is as well, hence the structure of J can be deduced from a small part of it. Along these same lines, Theorem 11.4 can be used to prove a surprising result about the set J:

Theorem 11.5 *The set J for a given rational function $\phi(z)$ contains an open disc D if and only if $J = \bar{\mathbb{C}}$.*

One way to see this is as follows.[8] Let D be an open disc contained in the Julia set of a function $\phi(z)$. Since it is also a neighborhood of a point z in J, it corresponds to the neighborhood N in the statement of Theorem 11.4. Any point $w \in \bar{\mathbb{C}} - \mathcal{E}$ is a closed set, thus Theorem 11.4 asserts the existence of a positive integer M such that $w \in \phi^M[D]$. Since $D \subset J$ the invariance of J under $\phi(z)$ implies that $\phi^M[D] \subset J$, hence the point w is in J. This argument can be applied to all non-exceptional points, and it therefore follows that $\bar{\mathbb{C}} - \mathcal{E} \subset J$. Since the exceptional set $\mathcal{E}$ contains at most two points, this implies that J equals the sphere except for at most two points. But J is perfect, and therefore closed, so $J = \bar{\mathbb{C}}$.

An example where the Julia set is the entire extended plane occurs with the function

$$l(z) = \frac{(z^2+1)^2}{4z(z^2-1)}, \tag{11.5}$$

which was discussed by Lattès in his note [1918a] dated January 7, 1918, and is known as Lattès' function.

Lattès' discovery of the function $l(z)$ was prompted by his plan, discussed in the previous chapter, to examine the Poincaré equation $F(su) = \phi(F(u))$ via the investigation of functions $F(u)$ for which $F(su)$ is a rational function of $F(u)$. For $s = 2$ the Weierstrass $\wp$-function is one such function, and Lattès observed that if the standard $\wp$-function coefficients g_2 and g_3 are fixed as $g_2 = 4$ and $g_3 = 0$, $\wp(u)$ satisfies the Poincaré equation

$$\wp(2u) = l(\wp(u)),$$

where $l(z)$ is Lattès' function.

Both Fatou and Julia later showed that the Julia set of $l(z)$ is the extended plane, but the peculiarities of this function were first noticed by Lattès, who took note of what he considered strange behavior, namely, that for fixed z, the perfect component of the closure of the set

$$\{\wp(2^n u)\} = \{l^n(\wp(u))\},$$

which is the forward orbit of a point z satisfying $z = \wp(u)$, is generally a particular closed curve which varies with the choice of z.

Julia, however, quickly realized the significance of Lattès' function. Writing three weeks later in the *Compte rendu* of January 28, 1918, he observed that the Julia set of $l(z)$ is the entire extended plane, and therefore $l(z)$ had no attracting periodic orbits. Thus, the curves Lattès noticed are explained by the fact that for arbitrary z, $\{l^n(z)\}$ never converges to an attracting orbit but rather meanders through the plane. As Julia observed, "The oddities Lattès noticed thus seem less surprising ... [1918b:153]."

---

[8] **Julia and Fatou proved this theorem directly from Theorem 11.3.**

11.6 Iteration on the Fatou Set

As in the examples described in the previous sections, the Julia set J partitions its complement, the Fatou set F, into a number of connected components on which the family

$$\mathcal{G} = \{\phi^n : n \in \mathbb{Z}_0^+\}$$

is normal. Consequently, the structure of F can be inferred from that of J. For example, if J is a TDP set, an infinite number of discrete continuous arcs, or a line segment, then F consists of a single component which is simply connected in the last case and infinitely connected in the first two. If the Julia set for $\phi(z)$ is a simple, closed curve then F consists of two simply connected components. If $J = \bar{\mathbb{C}}$, then F is empty. Finally, if J is a closed continuous curve with an infinite number of double points, then F consists of infinitely many simply connected components.

Montel's theory of normality is no less useful on the Fatou set than it was on the Julia set. In a manner that will be made precise over the next few paragraphs, Julia used Montel's theory of normal families to show that all points in a particular component of F exhibit similar behavior under iteration. In the process of doing this he extended Koenigs' local theory of iteration, which heretofore only described iteration on a disc D surrounding an attracting fixed point x, to the entire component of F which contained x. Although I will be outlining Julia's characterization of F, Fatou's view of F did not differ in any substantive way. However, Julia's conception of the Fatou set was expressed with greater clarity and precision.

The key to Julia's description of F is that the limit functions of the family $\mathcal{G}$ on F correspond to the limit points in the forward orbit of points z_0 in F under $\phi(z)$, that is, the set

$$O^+(z_0) = \{\phi^n(z_0)\}.$$

For example, suppose that w_0 is a subsequential limit point of the sequence $O^+(z_0)$ for some z_0 in F. There then exists a subsequence of points $\{\phi^{n_i}(z_0)\}$ which converges to w_0. At the same time, the normality of $\mathcal{G}$ on F implies that the corresponding sequence of functions $\{\phi^{n_i}\}$ converges uniformly on a neighborhood D of z_0 to a function $G(z)$ satisfying $G(z_0) = w_0$.

Conversely, given a point z_0 from a subdomain D of F and a limit function $G(z)$ from $\mathcal{G}$ which is defined on D, infinitely many iterates $\phi^n(z_0)$ are arbitrarily close to the point $G(z_0)$.

The following theorem of Montel enabled Julia to use this correspondence between the limit functions of $\mathcal{G}$ and the limit points of $O^+(z_0)$ to characterize iteration on F (Theorem 8.6 above):

Theorem 11.6 (Montel) *Let $\{f_i\}$ be a sequence of functions from a meromorphic family $\mathcal{F}$ which is normal on a domain A. Suppose that the f_i converge to a function $G(z)$ on an infinite set of points which has a limit point in A. Then the f_i converge uniformly to $G(z)$ on A.*

Theorem 11.6 is particularly useful in the event that a component A of F contains an attracting fixed point of x of $\phi(z)$. Koenigs showed in his paper [1884] that there exists a disc D surrounding x on which the iterates of $\phi(z)$ converge uniformly on D to x, that is, for all z in D

$$\lim_{n\to\infty} \phi^n(z) = x.$$

Viewed from the perspective of normal families, this implies that $\mathcal{G}$ has a unique limit function $G(z) \equiv x$ on D. The above theorem of Montel's in turn implies that the convergence of $\mathcal{G}$ to $G(z) \equiv x$ can be extended from D throughout A, since $\mathcal{G}$ is normal on A. Consequently, due to the correspondence between limit functions of $\mathcal{G}$ and limit points of the sequence of $\{\phi^n(z_0)\}$, all points in A converge to x under iteration. This behavior is summarized in the following theorem.

Theorem 11.7 *Let $\phi(z)$ have a fixed point x satisfying $|\phi'(x)| < 1$. Let A be the component of F which contains x. Then all z in A converge to x under iteration by $\phi(z)$.*

The preceding extends easily to components of F which do not contain an attracting fixed point. For example, if instead of converging to a fixed point x on D, $\phi^n(z_0)$ converges to a period p orbit $\{x_0, \ldots, x_{p-1}\}$, then the limit functions on D of the family $\mathcal{G}$ are the functions $G_i(z) \equiv x_i$, corresponding to subsequences of $\{\phi^n(z_0)\}$ which converge to x_i. The convergence of the limit functions $G_i(z)$ on D can then be extended to the entire component of F which contains D. Likewise, if all points in an attracting petal A_i given by the Flower Theorem (see Theorem 5.1) converge to a fixed point x of a function $\phi(z)$ satisfying $\phi'(x) = 1$, this behavior extends to the component of the Fatou set F which contains A_i. Julia thus showed that on any component A from the Fatou set, iterates of $\phi(z)$ exhibit the same behavior throughout A.

Julia summed up his discussion by observing that Montel's Theorem provides a concise means of extending Koenigs' local results to the entire component of F which contains x:

> To say that if in A the behavior of the sequence $\phi(z), \phi^2(z), \ldots, \phi^n(z)$, ... is the same, is to say that any subsequence converges on all areas interior to A or on none of these areas.
>
> This is the manner in which the perfect set J delimits the diverse regions of convergence of the plane: in all connected regions bounded by J, the character of the sequence $\phi(z), \phi^2(z), \ldots, \phi^n(z), \ldots$ is the same.
>
> The preceding theorem [i.e., Theorem 11.6] makes a genuine bridge between the general study of iteration in the entire plane and the local study, which has been the sole enterprise [of the study of iteration] up to this point [1918:195–96].

Julia and Fatou also studied the limit functions of $\mathcal{G}$ on components of F which have the odd property that there exists no open disc D on which the iterates converge to a period p orbit, where $p \geq 1$. Fatou called such components *singular domains*. Examples of singular domains are very difficult to construct, so much so, that neither Julia nor Fatou were able to construct one, or even prove their existence. Nonetheless, if a singular domain S exists, and if it contains a disc D on which the sequence $\{\phi^{n_i}\}$ converges to a limit function $G(z)$, Julia observed that Theorem 11.6 implies that the convergence to $G(z)$ can be extended to A. Subsequent to these works of Fatou and Julia singular domains have been shown to exist. More will be said about this subject below.

11.7 A Limit on the Number of Attracting Orbits

After successfully using Montel's theory of normal families to partition the sphere into a number of components such that on each component the iterates of $\phi(z)$ exhibit the same behavior, Fatou and Julia inspected this behavior more closely. One of the most important results they proved was that the number of attracting and neutral periodic points is finite.[9]

This result resolved a long-standing question of Koenigs', which he posed at the end of his paper [1884], as to whether there were infinitely many attracting periodic points. Koenigs feared that there might be, in which case there would be infinitely many attracting domains. This would in turn make his proposed division of the plane into regions of attraction, which he already suspected might be an impossible task, all the more difficult.To a certain extent, Koenigs' fears were justified because although the number of attracting domains is finite, the number of components of the Fatou set need not be. In this section, I will discuss the discovery that the number of attracting periodic points is finite; in the next I will show how Fatou and Julia demonstrated that the Fatou set frequently has infinitely many components.

The discovery that the number of attracting orbits is finite was achieved by paying careful attention to a set that previous to Fatou's 1906 note had been largely ignored, namely, the behavior of the forward orbit of the critical points of $\phi(z)$. Actually, Fatou looked at a related set in Theorem 7.1 from his note [1906a], the orbit of the critical values. A critical *point* c of $\phi(z)$ satisfies $\phi'(c) = 0$, perhaps under a local coordinate change, and a critical *value* is the image of a critical point. In order to discuss these results with precision, it will be helpful to distinguish the components of F which contain an attracting orbit from those which do not.

Definition 11.8 *The* immediate domain of attraction *for an attracting fixed point x of the function $\phi(z)$ is the component of F which contains x. This set is also*

[9] A *neutral* periodic point is one in which the multiplier is equal to one in modulus. A *rationally neutral* periodic point is one in which the multiplier is a root of unity, and an *irrationally neutral* periodic point is one in which the multiplier is one in modulus but is not a root of unity.

sometimes called the immediate basin of attraction *of* x. *It is denoted* A_x, *and components of* F *which also converge to* x *under iteration but do not contain* x *are called* preperiodic. *The complete set of points in* $\bar{\mathbb{C}}$ *which converge to* x *under iteration by* $\phi(z)$ *is called the* total domain of attraction *of* x *or the* basin of attraction *of* x. *More generally, the immediate domain of attraction of a periodic orbit is the union of the* p *components* A_i *of* F *which contain an* x_i. *Preperiodic domains are those components which converge to the periodic orbit but which do not contain any of the* x_i. *The total domain of convergence is the totality of points which converge to the periodic orbit* $\{x_0, \ldots, x_{p-1}\}$.

Fatou [1920:61] and Julia [1918:203] each proved the following theorem and corollary which serves to limit the number of attracting periodic orbits.

Theorem 11.9 *The immediate domain of attraction for a periodic orbit contains a critical point of the function* $\phi(z)$.

Both Fatou and Julia showed that a degree d rational function has at most $2d-2$ critical points, which makes the following immediate.

Corollary 11.10 *The number of attracting periodic orbits is less than or equal to the number of critical points of the function* $\phi(z)$, *which is at most* $2d-2$, *where* d *is the degree of* $(\phi(z))$.

Both Fatou [1920a:63] and Julia [1918:211] extended Theorem 11.9 to rationally neutral orbits (periodic points whose multiplier is a root of unity) and proved the following:

Theorem 11.11 *The number of attracting orbits plus the number of rationally neutral orbits is bounded by the number* $2d-2$, *where* d *is the degree of* $\phi(z)$.

The attempts of Fatou and Julia to find an upper bound on neutral domains differed in one substantive aspect. Both hypothesized that singular domains S might exist, but could not prove that they existed if the degree of $\phi(z)$ was strictly greater than one.[10] Both correctly recognized that if such domains existed, then iteration of $\phi(z)$ in S behaves like the irrational rotation $z \mapsto$ to $e^{i\theta}z$, where $\theta/\pi \notin \mathbf{Q}$. Only Fatou, however, sought to find an upper bound on the number of singular domains, and in fact proved that should singular domains exist, their number would be at most $4d-4$ [1920a:69].

Aside from the fact that Julia did not seek a bound on the number of singular domains, his study of these domains was quite thorough [1918:311ff]. He related the existence of singular domains, which he called centers, to the existence of an irrationally neutral fixed point x of $\phi(z)$ in S satisfying $\phi'(x) = e^{i\theta}$ where $\theta/\pi \notin \mathbf{Q}$.[11]

[10] Singular domains are components of F which converge neither to attracting nor to rationally neutral domains. Sometimes these domains are called rotation domains.

[11] It has been shown since that S can be annular and therefore need not contain a fixed point. Iteration then acts like an irrational rotation of an annulus.

In his only application of functional equations in [1918], Julia showed that if such a singular domain S exists, then there also exists an analytic solution $F(z)$ to the following Schröder functional equation

$$F(\phi(z)) = e^{i\theta} F(z) \tag{11.6}$$

throughout S, and furthermore, that x is in the Fatou set F. On the other hand, if an irrationally neutral fixed point did not correspond to a singular domain, there exists no neighborhood of x on which equation (11.6) could be solved, and x is a member of the Julia set J.

Perhaps the reason that Julia sought no bound on the number of singular domains in [1918] is that at the time he wrote [1918] he believed, incorrectly as it turns out, that singular domains did not exist if the degree of $\phi(z)$ is strictly greater than 1. That Julia thought this is indicated in his note [1919] in which he announced that he had proved that singular domains did not exist and that all irrationally neutral fixed points were therefore in the Julia set [1919:147].

He was mistaken in this claim, for in his paper [1942] Carl Siegel proved that there exist simply connected components of F containing an irrationally neutral fixed point x in which iteration of $\phi(z)$ acts like an irrational rotation provided that $\phi'(x)$ satisfies a certain number theoretic condition (see Theorem 5.4).

Later, Michael Herman (1942-) showed in his paper [1979] that annular domains also exist on which iteration by $\phi(z)$ acts like an irrational rotation. Dennis Sullivan (1941-) subsequently showed in his papers [1983] and [1985] that these are the only kinds of singular domains which can exist if $\phi(z)$ is a rational function, and consequently components of the Fatou set are of one of the following types: they converge either to attracting or periodic orbits, they converge to a rational neutral periodic orbit or they are rotation domains. Mitsuhiro Shishikura (1960-) in his paper [1987] showed that the number of attracting orbits plus the number of rationally neutral orbits plus the number of rotation domains is at most $2d-2$.

In any event, Julia, in his *Notice sur les travaux scientifiques*, written in anticipation of his election to the French Academy of Sciences in 1934, noted that he had, in the argument of the proof referred to in his note [1919], made false assumptions concerning the boundary points of components in the Fatou set, which he observed threw doubt on his earlier assertion that singular domains could not exist [1968, Volume I:22].

Fatou and Julia were not the first mathematicians to address the issue of the existence or non-existence of solutions to the equation

$$F(\phi(z)) = \phi'(x)F(z) \tag{11.7}$$

where $\phi(z)$ is analytic around a fixed point x satisfying $\phi'(x) = e^{i\theta}$, where θ is not commensurate with π. In 1915 the American mathematician George Pfeiffer related solutions to the above Schröder equation to the existence of a conformal mapping of a curvilinear angle of angle θ to a rectilinear angle of the same magnitude.

Pfeiffer found no solutions to (11.7), but he did prove two interesting facts. The first was the existence of analytic functions for which no solutions to (11.7) exist. The second was that if $g(z)$ is any analytic function with a fixed point at x satisfying $|g'(x)| = 1$, then a perturbation $\phi(z)$ of the function $g(z)$ exists such that equation (11.7) has no solutions. The function $\phi(z)$ is a perturbation of $g(z)$ in the sense that for a given $\epsilon > 0$, there exists a function $\phi(z)$ such that $\phi(x) = x$, $\phi'(x)$ is one in modulus but not a root of unity, and $|\phi_i - g_i| < \epsilon$, where ϕ_i and g_i are the corresponding coefficients of the Taylor expansions for $\phi(z)$ and $g(z)$ about x.

Pfeiffer attributed his awareness of the connection between equation (11.7) and the conformal mapping of curvilinear angles to a series of lectures given at Columbia University by the American mathematician Edward Kasner, who was Ritt's thesis advisor.

Neither Fatou nor Julia gave any indication that they were aware of the work of Pfeiffer, Kasner or Ritt.

11.8 The Number of Components of the Fatou Set

Julia and Fatou each offered examples of Julia sets consisting of a continuous curve which crosses itself infinitely often. Since J can not have any interior points unless it equals $\bar{\mathbb{C}}$, this implies that the curve partitions F into infinitely many components. However, the number of attracting, neutral or singular orbits is finite, and therefore, at least one of these orbits must have infinitely many preperiodic components (see Definition 11.8).

This situation is not at all unusual. Indeed, the following theorem, which Julia [1918:221ff] and Fatou [1920a:50ff] each proved, justifies Koenigs' fears that division of the plane into regions is extremely complicated.

Theorem 11.12 *If F does not have an infinite number of components it has at most two.*

To prove this theorem, both Fatou and Julia relied on two major lemmas, the first of which, explains why F can have infinitely many components.

Lemma 11.13 *If there exists at least one component of F which is preperiodic, then there are an infinite number of preperiodic domains.*

I will sketch Fatou's proof in the case where the domain of attraction for an attracting fixed point x has a preperiodic component. Fatou's argument typifies a kind of topological argument he employed repeatedly.

Let A_x be the immediate domain of attraction of a fixed point x. Fatou had previously shown that $\phi(z)$ is forward invariant on A_x, that is, it satisfies $\phi[A_x] \subset A_x$. This follows easily from the connectedness of A_x.

To see that there cannot be a unique preperiodic component of A_x, suppose one exists. Call it A' and pick $z' \in A'$ such that $\phi(z') = z \in A_x$. Since all points in A'

converge to A_x, such a point exists. The function $\phi(z)$ is rational, thus it is also an onto map of $\bar{\mathbb{C}}$, hence z' has a preimage z'', which due to the forward invariance of A_x must be in A'. The point $\phi(z'') = z'$ converges to x under iteration, so z'' must as well. Both z' and z'' are in A', so there exists a continuous curve γ in A' connecting z' to z''. Since A and A' are distinct components of F they have no points in common and are separated by the Julia set J. The continuity of $\phi(z)$ therefore implies that $\phi[\gamma]$ is either entirely in A or in A'. But $\phi(z'') = z'$ is in A', and $\phi(z') = z$ is in A_x, so this can not be. The contradiction is resolved only if there is a second preperiodic component A'' containing z'' such that $\phi[A''] \subset A'$. Applying the above argument to A'', there must be a third preperiodic component, and so forth.

The second lemma used in the proof of Theorem 11.12 is the following.

Lemma 11.14 *Let $\phi(z)$ be a rational function of degree $d \geq 2$. A completely invariant component A is defined to be one satisfying both $\phi[A] = A$ and $\phi^{-1}[A] = A$ where $\phi^{-1}(x)$ is the total inverse of $\phi(z)$, that is, $\phi^{-1}(z)$ is the complete set of inverse images of z. A completely invariant, simply connected component A of F has $d - 1$ critical points, counted with multiplicity.*

Lemma 11.14 implies that there can be no more than two simply connected, completely invariant components of F. The existence of three such components in turn implies the existence of at least $3d - 3$ critical points, which contradicts the corollary to Theorem 11.9. Fatou and Julia also showed that there can not be three or more completely invariant domains, since the existence of three completely invariant domains implies that they are all simply connected.

Theorem 11.12 now follows readily: if there are three or more components then Lemma 11.14 implies that at least one is not completely invariant, hence it is preperiodic, in which case Lemma 11.13 asserts that there must be infinitely many other preperiodic components. Therefore, if the number of components of F is finite, there can be no more than two.

11.9 Newton's Method Again

As recounted in the opening chapter, the study of Newton's method for complex functions was the principal motivation for the first important work concerning the iteration of complex functions, Schröder's papers [1870] and [1871]. The question of the convergence of Newton's method for complex polynomials also interested Cayley, and he published several short works concerning Newton's method, the most important of which was his paper [1879a]. Although their respective approaches were quite different, Cayley and Schröder each proved that for the complex polynomial $q(z) = z^2 - 1$ the Newton's method function for $q(z)$, $N_q(z) = (z^2 + 1)/2z$, converges to -1 on the left half-plane, to 1 on the right half-plane, and to neither root on the imaginary axis. The Newton's method function for the general quadratic exhibits analogous behavior, as both Schröder and Cayley indicated.

It is important to keep in mind that the methods each man used were ad hoc in the sense that they relied on techniques which did not generalize to the study of Newton's method for higher degree polynomials. It is therefore no surprise that, aside from Schröder's attracting fixed point theorem which shows that the roots of an arbitrary function $f(z)$ are attracting fixed points of the Newton's method function for $f(z)$, neither man had any success in extending their arguments from the quadratic case to higher degree polynomials.

The work of Fatou and Julia, however, does away with many of the difficulties which frustrated Schröder and Cayley. For example, Fatou gave an example of Newton's method for the cubic $c(z) = z^3 - 1$, in which he showed that the Julia set was a curve which divided the plane into infinitely many components [1920a:89–90]. That the Julia set for the Newton's method function $N_c(z)$ should be such a curve follows immediately from the fact that, as Schröder's fixed point theorem explains, the three roots of $c(z)$ are attracting fixed points of $N_c(z)$. Since each attracting fixed point corresponds to at least one component of the Fatou set, there are at least three such components. Consequently, Theorem 11.12 asserts that there are infinitely many components to the Fatou set.

Julia discussed Newton's method in considerable detail [1918:249ff] and gave two reasons why those who had previously studied Newton's method had run into difficulties. The first has to do with the fact that the roots of a polynomial $p(z)$ of degree n are also critical points for the corresponding Newton's method function, $N_p(z)$. Since all told $N_p(z)$ has at most $2n - 2$ critical points (the corollary to Theorem 11.9), and since any completely invariant component of the Fatou set F for $N_p(z)$ must have $n - 1$ critical points (Lemma 11.14), it follows that whenever $n > 2$, there is at most one root for which the corresponding attracting domain consists of a single, completely invariant component of F. Thus, the corresponding domains of attraction for the other $n - 1$ roots of $p(z)$ must have infinitely many preperiodic components.

After outlining these facts, Julia observed:

> In general, the actual division of the plane into regions, each of which converge to a determined root of $p(z) = 0$, will be an impractical problem, since at least $n - 1$ of the roots have a domain of convergence consisting of infinitely many areas, hence one must divide the plane into infinitely many regions. Here is the reason that Cayley's attempt to apply Newton's method to equations of degree ≥ 3 was checked [1918:254].

The second inherent difficulty in the study of Newton's method which Julia observed involves the fact that the Julia set for a rational function is a subset of the closure of the backward orbit of any non-exceptional point in $\bar{\mathbb{C}}$ (see Theorem 11.3). This implies that contained in any neighborhood of a point in the Julia set of $N_p(z)$ are points which converge to each of the attracting fixed points of $N_p(z)$, which makes the direct calculation of the Newton's method function very difficult, since the behavior under iteration by $N_p(z)$ of points arbitrarily close to one another could vary greatly.

Julia's omission of any reference to Schröder in the above quotation is telling, since it underscores how far Schröder's work in iteration had faded into obscurity. Although Fatou and Julia were probably aware of Schröder, for example, both used the term Schröder functional equation and would have certainly heard of him through the work of Koenigs and Leau, it is not clear how much they actually knew about his contributions. Neither Fatou nor Julia discussed Schröder's accomplishments in any detail, much less his study of Newton's method. Indeed, each credited Cayley with establishing the convergence properties of Newton's method for the quadratic.

This is especially ironic since Cayley's paper was conceived along narrow lines and consequently contributed relatively little to the general study of iteration, while Schröder's study was founded upon the general principles. In many respects, the basic question Schröder sought to answer is the same one which motivated Fatou and Julia, namely, given a point in the plane, what are its properties under iteration by a given function $\phi(z)$.

In any event, I hope that viewing the investigations of Cayley and Schröder from the point of view of Fatou and Julia not only illuminates the difficulties the former faced, but also emphasizes the power of the theory developed by Fatou and Julia.

Bibliography

[1881] Abel, Niels H. *Oeuvres complètes d'Abel*, 2 volumes, Johnson Reprint Corporation, 1964.

[1823] Abel, Niels H. "Méthod Générle pour trouver des fonctions d'une seule quantité variable, lorsqu'une propriété de ces fonctions est exprimée par une équation entre deux variables." *Magazin for Naturvidenskaberne*, 1823. Also in [1881,I:1–10].

[1824?] Abel, Niels H. "Determination d'une fonction au moyen d'une équation ne contient qu'une variable." Also in [1881,II:36–39].

[1826a] Abel, Niels H. "Recherche des fonctions de deux quantités variable indépendentes x et y, telles que $f(x,y)$, qui ont la propriété que $f(z,f(x,y))$ est une fonction symétrique de z, x et y." *Journal für die reine und angewandte Mathematik*, **1**, 1826. Also in [1881,I:61–65].

[1826b] Abel, Niels H. "Sur la fonction $\psi x = x + \frac{x^2}{2^2} + \frac{x^3}{3^2} + \cdots + \frac{x^n}{n^2} + \cdots$." *Journal für die reine und angewandte Mathematik*, **1**, 1826. Also in [1881,I:219-50].

[1826c] Abel, Niels H. "Note sur la fonction $\psi x = x + \frac{x^2}{2^2} + \frac{x^3}{3^2} + \cdots + \frac{x^n}{n^2} + \cdots$." Unpublished, 1826. Also in [1881,I:189-93].

[1827] Abel, Niels H. "Sur les fonctions qui satisfont a l'équation $\phi x + \phi y = \psi(xfy + yfx)$." *Journal für die reine und angewandte Mathematik*, **2**, 1827. Also in [1881,I:389-98].

[1966] Aczél, János. *Lectures on Functional Equations and their Applications.* Translated by Scripta Technica Inc., Academic Press, 1966.

[1989] Aczél, János and Dhombres, Jean. *Functional Equations in Several Variables, Volume 31, Encylopedia of Mathematics and its Applications*, Cambridge University Press, 1989.

[1891] Appell, Paul. "Sur des équations différentielles linéaires transformables en elles-mêmes par un changement de fonction et de variable." *Acta Mathematica*, **15**, 1891, 281–315.

[1883] Ascoli, Giuol. "La curve limite di una varietà data di curve." *Atti della Reale Accademia dei Lincei*, Series 3, **18**, 1883, 521–86.

[1889] Arzelà, Cesare. "Funzioni di linee." *Atti della Reale Accademia dei Lincei, Rendecotti*, Series 4, **5**, 1889, 342–48.

[1895] Arzelà, Cesare. "Sulle funzioni di linee." *Memoire Accademia della scienze dell'instituto di Bologna*, Series 5, **5**, 1895, 255–244.

[1896] Arzelà, Cesare. "Sul principio di Dirichlet." *Rendiconto delle sessioni dell Accademia della scienze dell'instituto di Bologna*, **1**, 1896–97, 71–84.

[1899] Arzelà, Cesare. "Sulle serie di funzioni." *Memoire Accademia della scienze dell'instituto di Bologna*, Series 5, **8**, 1899-90, 131–86.

[1900] Arzelà, Cesare. "Sulle serie di funzioni." *Memoire Accademia della scienze dell'instituto di Bologna*, Series 5, **8**, 1899-90, 701–44.

[1820] Babbage, Charles. *Examples of the Solutions of Functional Equations*, 1820.

[1897] Baire, René. "Sur le théorie générale des fonctions de variables réelles." *Comptes rendus hebdomadaires des Séances de l'Académie des Sciences*, **125**, 1897, 691–94.

[1898a] Baire, René. "Sur les fonctions discontinues développables en séries de fonctions continues." *Comptes rendus hebdomadaires des Séances de l'Académie des Sciences*, **126**, 1898, 884–87.

[1898b] Baire, René. "Sur les fonctions discontinues qui se rattachent aux fonctions continues." *Comptes rendus hebdomadaires des Séances de l'Académie des Sciences*, **126**, 1898,1621–23.

[1899] Baire, René. "Sur les fonctions de variables réelles." *Annali di Matematica pura ed applicata*, Series 3, **3**, 1899, 1–123.

[1980] Bashmakova, I. G. "Egor Zolotarev" from *The Dictionary of Scientific Biography*, Charles Schribners and Sons, Volume 14, 1980.

[1991] Beardon, Alan F. *Iteration of Rational Functions*, Springer-Verlag, New York, 1991.

[1883] Bendixson, Ivar. "Quelques thérèmes de la theoreie des ensembles." *Acta Mathematica*, **2**, 415–29.

[1915] Bennett, Albert A. "The Iteration of Functions of One Variable." *Annals of Mathematics*, **17**, 1915–16, 23–60.

[1915] Bennet, Albert A. "A Case of Iteration of Several Variables." *Annals of Mathematics*, **17**, 1915–16, 180–96.

[1984] Blanchard, Paul. "Complex Analytic Dynamics on the Riemann Sphere." *Bulletin of the American Mathematical Society*, **11**, 1984, 85–141.

[1896] Borel, Émile. "Démonstration élémentaire d'un thérème de M. Picard sur les fonctions entières." *Comptes rendus hebdomadaires des Séances de l'Académie des Sciences*, **122**, 1896, 1045–48.

[1898] Borel, Émile. *Leçons sur la théroie des fonctions*, Gauthier-Villars, 1898.

[1900] Borel, Émile. *Leçons sur les fonctions entières*, Gauthier-Villars, 1900.

[1899] Böttcher, Lucyan. *Beiträge zu der Theorie der Iterationsrechnung*, Oswald Schmidt, Leipzig, 1898.

[1899] Böttcher, Lucyan. "Zasady Rachunku Iteracyjnego (Principles of Iteration)." *Prace matematyczno-fizyczne, Warsaw*, **10**, 1899-1900, 65–101.

[1900] Böttcher, Lucyan. "Równania funkcyjne podstawnicze (Functional Substitution Equations)." *Wiadomości matematyczne, Warsaw*, **4**, 1900, 233–35.

[1901] Böttcher, Lucyan. "O wianościach pewnych wyznaczników funkcyjnych (A Lemma Concerning the Grévy Determinant)." *Kraków akademij umiejetnosci, Wydzai matematyczno-przyrodniczy (Proceedings of the Academia of Sciences, Mathematics and Natural Science, Krakow)*, **18**, 1901, 382–89.

[1904] Böttcher, Lucyan. "The principal laws of convergence of iterates and their application to analysis (Russian)." *Bulletin Kasan Mathematical Society*, **14**, 1904, 155–234.

[1897a] Bourlet, Carlo. "Sur les opérations en général et les équations différentielles linéaires d'ordre infini." *Annales Scientifiques de l'École Normale Supérieure*, **14**, 1897, 133–90.

[1897b] Bourlet, Carlo. "Sur les opérations en général." *Comptes rendus hebdomadaires des Séances de l'Académie des Sciences*, **124**, 1897, 348–51.

[1897c] Bourlet, Carlo. "Sur certaines équations différentielles." *Comptes rendus hebdomadaires des Séances de l'Académie des Sciences*, **124**, 1897, 1431–44.

[1898] Bourlet, Carlo. "Sur l'itération." *Comptes rendus hebdomadaires des Séances de l'Académie des Sciences*, **126**, 1898, 583–85.

[1899a] Bourlet, Carlo. "Sur certaines équations analogues aux équations différentielles." *Annales Scientifiques de l'École Normale Supérieure*, **16**, 1899, 333–75.

[1899b] Bourlet, Carlo. "Sur le problème de l'itération." *Annales de la Faculté des Sciences de l'Université de Toulouse*, **12**, 1899, c1–c12.

[1921] Buhl, A. "Éloge de Samuel Lattès." *Extrait des Mémoires de L'Académie des Science*, Volume IX, 1921, 1–13.

[1931] Buhl, A. "Gabriel Koenigs." *L'Enseignement Mathématique*, **30**, 1931, 286–87.

[1911] Cajori, Florian. "Historical Note on the Newton-Raphson Method." *American Mathematical Monthly*, **18**, 1911, 29-33.

[1980] Cajori, Florian. *A History of Mathematics*, third edition, Chelsea Publishing Company, 1980.

[1879] Cantor, Georg. "Ueber unendliche, linear Punktmannichfaltigkeiten," Part 1. *Mathematische Annalen*, **15**, 1879, 1–7.

[1883] Cantor, Georg. "Ueber unendliche, linear Punktmannichfaltigkeiten," Part 4. *Mathematische Annalen*, **21**, 1883, 51–58, 545–91.

[1821] Cauchy, Augustin-Louis. *Cours d'analyse de l'Ècole Royale Polytechnique.* Also in *Oeuvres complète*, Series 2, Tome 3.

[1897] Cayley, Arthur. *The Collected Mathematical Papers of Arthur Cayley*, Cambridge University Press, 1889-1897.

[1860] Cayley, Arthur. "On Some Numerical Expansions." *Quarterly Journal of Pure and Applied Mathematics*, **3**, 1860, 366–69. Also in [1897,IV:470–72].

[1879a] Cayley, Arthur. "Applications of the Newton–Fourier Method to an Imaginary Root of an Equation." *Quarterly Journal of Pure and Applied Mathematics*, **16**, 1879, 179-85. Also in [1897,XI:114–121].

[1879b] Cayley, Arthur. "The Newton–Fourier Imaginary Problem." *American Journal of Mathematics*, **2**, 1879, 97. Also in [1897,X:405].

[1880] Cayley, Arthur. "On the Newton–Fourier Imaginary Problem." *Proceedings of the Cambridge Philosophical Society*, **3**, 1880, 231–32. Also in [1897,XI:143].

[1890] Cayley, Arthur. "Sur les racines d'une èquation algèbrique." *Comptes rendus hebdomadaires des Séances de l'Académie des Sciences*, **110**, 1890, 174–76 and 215–18. Also in [1897,XIII:33–42].

[1990] Chabert, Jean-Luc. "Un demi-siecle de fractales: 1870–1920." *Historia Mathematica*, **17**, 1990, 339-365.

[1872] Darboux, Gaston. "Sur un théorème relatif à la continuité des fonctions." *Bulletin des Sciences mathématiques et astronomiques*, **3**, 1872, 307–313.

[1875] Darboux, Gaston. "Mémoire sur les fonctions discontinues." *Annales Scientifiques de l'École Normale Supérieure*, Series 2, **4**, 1875, 58–112.

[1879] Darboux, Gaston. "Addition au mémoire sur les fonctions discontinues." *Annales Scientifiques de l'École Normale Supérieure*, Series 2, **8**, 1879, 195–201.

[1881] Darboux, Gaston. Review of Abel's *Oeuvres*. *Bulletin des Sciences mathematiques et astronomiques*, Series 2, **5**, Part 1, 1881, 457–62.

[1989] Devaney, Robert. *An Introduction to Chaotic Dynamical Systems*, second edition. Addison-Wesley Publishing Company, 1989.

[1982] Dhombres, J. "On the historical role of functional equations." *aequationes mathematicae*, **25**, 293–99.

[1981] Dieudonne, Jean. *History of Functional Analysis*. North-Holland Publishing Company, Amsterdam, 1981.

[1990] Dieudonne, Jean. "Paul Montel from *The Dictionary of Scientific Biography*, Charles Schribners and Sons, Volume 18, Supplement II, 1990, 649-500.

[1878] Dini, Ulisse. *Sirie di Fourier e altre rappresentazioni analitiche delle funzioni di una variabile reale*, Pisa, 1878.

[1880] du Bois-Reymond, Paul. "Der Beweis des Fundamentalsatzes der Intergralrechnung: $\int_a^b F'(x)dx = F(b) - F(a)$." *Mathematische Annalen*, **16**, 1880, 115–28.

[1884] Farkas, Jules. "Sur les fonctions itératives." *Journal de Mathématics Pure et Appliquées*, Series 3, **10**, 1884, 101–08.

[1976] Dugac, Paul. "Notes et documents sur le vie et l'oeuvre de René Baire." *Archive for History of Exact Sciences*, **15**, 1976, 289-383.

[1960/61] Erdös, Paul and Jobitinsky, E. "On Analytic Iteration." *Journal d'analyse mathematique*, **8**, 1960/61, 361–76.

[1906a] Fatou, Pierre. "Sur les solutions uniformes de certaines équations fonctionnelles." *Comptes rendus hebdomadaires des Séances de l'Académie des Sciences*, **143**, 1906, 546–48.

[1906b] Fatou, Pierre. "Séries trigonométriques et séries de Taylor." *Acta Mathematica*, **30**, 1906, 335–400.

[1910] Fatou, Pierre. "Sur une classe remarquable de séries de Taylor." *Annales Scientifiques de l'École Normale Supérieure*, Series 3, **27**, 1910, 43–53.

[1913] Fatou, Pierre. "Sur les lignes singulières des fonctions analytiques." *Bulletin de la Société mathematique de France*, **41**, 1913, 113–19.

[1917a] Fatou, Pierre. "Sur les substitutions rationnelles." *Comptes rendus hebdomadaires des Séances de l'Académie des Sciences*, **164**, 1917, 806–08.

[1917b] Fatou, Pierre. "Sur les substitutions rationnelles." *Comptes rendus hebdomadaires des Séances de l'Académie des Sciences*, **165**, 1917, 992–95.

[1918a] Fatou, Pierre. "Sur les équations fonctionnelles et la propriétés de certaines frontières." *Comptes rendus hebdomadaires des Séances de l'Académie des Sciences*, **166**, 1918, 204–06.

[1918b] Fatou, Pierre. "Sur les suites de fonctions analytiques." *Comptes rendus hebdomadaires des Séances de l'Académie des Sciences*, **167**, 1918, 1024–26.

[1919] Fatou, Pierre. "Sur les équations fonctionelles." *Bulletin de la Société mathématique de France*, **47**, 1919, 161–271.

[1920a] Fatou, Pierre. "Sur les équations fonctionelles." *Bulletin de la Société mathématique de France*, **48**, 1920, 33–94.

[1920b] Fatou, Pierre. "Sur les équations fonctionelles." *Bulletin de la Société mathématique de France*, **48**, 1920, 208–314.

[1921a] Fatou, Pierre. "Sur les domaines d'existence de certaines fonctions uniformes." *Comptes rendus hebdomadaires des Séances de l'Académie des Sciences*, **173**, 1921, 344–46.

[1921b] Fatou, Pierre. "Sur les fonctions qui admettent plusieurs théorèmes de multiplication." *Comptes rendus hebdomadaires des Séances de l'Académie des Sciences*, **173**, 1921, 571–73.

[1921c] Fatou, Pierre. "Sur un groupe de substitutions algébriques." *Comptes rendus hebdomadaires des Séances de l'Académie des Sciences*, **173**, 1921, 694–66.

[1922] Fatou, Pierre. "Sur les équations fonctionnelles." Gauthier-Villars, 1922. This consists of the papers [1919], [1920a] and [1920b] published in book form under auspices of Société Mathématique.

[1922a] Fatou, Pierre. "Sur les fonctions méromorphes de deux variables." *Comptes rendus hebdomadaires des Séances de l'Académie des Sciences*, **175**, 1922, 862–65.

[1922b] Fatou, Pierre. "Sur certaines fonctions uniformes de deux variables." *Comptes rendus hebdomadaires des Séances de l'Académie des Sciences*, **175**, 1922, 1030–33.

[1922c] Fatou, Pierre. "Sur les l'itération des certains fonctions algébriques." *Bulletin des Sciences mathématiques et astronomiques*, **46**, 1922, 188–198.

[1922d] Fatou, Pierre. "Note sur les fonctions invariantes par une substitution rationelle." *Bulletin de la Société mathématiques de France*, **50**, 1922, 37–41.

[1923a] Fatou, Pierre. "Sur le frontières de certains domains." *Bulletin de la Société mathématiques de France*, **51**, 1923.

[1923b] Fatou, Pierre. "Sur l'itération analytiques et les substitutions permutables." *Journal de Mathématiques pures et appliquées*, Series 9, **2**, 1923, 343–438.

[1923c] Fatou, Pierre. "Sur les fonctions holomorphes et bornées à l'intérieur d'un cercle." *Bulletin de la Société mathématiques de France*, **51**, 1922.

[1924a] Fatou, Pierre. "Sur l'itération analytiques et les substitutions permutables." *Journal de Mathématiques pures et appliquées*, Series 9, **3**, 1924, 1–49.

[1924b] Fatou, Pierre. "Substitutions analytiques et équations fonctionnelles à deux variables." *Annales Scientifiques de l'École Normale Supérieure*, **41**, 1924, 67–142.

[1926] Fatou, Pierre. "Sur l'itération des fonctions transcendantes entières." *Acta Mathematica*, **47**, 1926, 337–370.

[1929] Fatou, Pierre. "Notice sur les travaux scientifiques," 1929.

[1875] Formenti, Carlo. "Su alcuni problemi di Abel." *Rendiconti, Instituto Lombardo di Scienze e Letter, Milan*, **8**, 1875, 276–82.

[1818] Fourier, Joseph. *Bulletin des Sciences par la Société philomatique de Paris*, 61, 1818.

[1831] Fourier, Joseph. *Analyses des equations déterminées*, 1831.

[1904] Fréchet, Maurice. "Généralisation d'un thérèome de Weierstrass." *Comptes rendus hebdomadaires des Séances de l'Académie des Sciences*, **139**, 1904, 848–50.

[1906] Fréchet, Maurice. "Sur quelques points du Calcul fonctionnel." *Rendiconti Circolo Matematico di Palermo*, 1906, 1–74.

[1953] Fréchet, Maurice. *Pages choisies d'analyse générale*, Gauthier-Villars, 1953.

[1978] Garnier, René. "Notice nécrologique sur Gaston Julia." *Comptes rendus hebdomadaires des Séances de l'Académie des Sciences, Vie Académique*, **286**, 1978, 126–33.

[1983] Gispert, Hélène. "Sur les fondements de l'analyse en France." *The Archive for History of Exact Sciences*, **28**, 1983, 37–106.

[1983] Gispert, Hélène. *La France mathématique: La Société mathématique (1872–1914)*, Société Française d'Histoire des Sciences et des Techniques, Paris, 1991.

[1892] Grévy, Auguste. "Sur les équations fonctionnelles." *Bulletin des Sciences mathématiques et astronomiques*, Series 2, **16**, Part 1, 1892, 311–13.

[1894] Grévy, Auguste. "Étude sur les équations fonctionnelles." *Annales Scientifiques de l'École Normale Supérieure*, Series 3, **11**, 1894, 249-323.

[1896] Grévy, Auguste. "Étude sur les équations fonctionnelles." *Annales Scientifiques de l'École Normale Supérieure*, Series 3, **13**, 1896, 295–338.

[1897] Grévy, Auguste. "Équations fonctionnelles avec second membre." *Bulletin de la Société mathématique de France*, **25**, 1897, 57–63.

[1901] Hadamard, Jacques. "Sur l'itération et les solutions asymptotiques des équations différentielles." *Bulletin de la Société mathematique de France*, **29**, 1901, 224–28.

[1881] Harnack, Axel. *Die Elemente der Differential- und Integral-rechnung*, B.G. Teubner, Leipzig.

[1975] Hawkins, Thomas. *Lebesgue's Theory of Integration*, Chelsea Publishing Company, 1975.

[1979] Herman, Michael. "Sur les conjugation différentiables des du cercle à les rotations." *IHES*, **49**, 5–233, 1979.

[1905] Hermite, Charles and Stieltjes, Thomas Jan. *Correspondance d'Hermite et de Stieltjes*, 2 volumes, Gauthier-Villars, Paris, 1905.

[1981] Hervé, Michel. "L'oeuvre de Gaston Julia." *Cahiers du Séminaire d'Histoire des Mathématiques*, **2**, 1981, 1–8.

[1992] Hervé, Michel. From *Supplement to The Dictionary of Scientific Biography*, Charles Schribners and Sons, preprint.

[1917] Humbert, George. "Sur une Communication de M. Gaston Julia, intitulée: Sur les substitions rationnells." *Comptes rendus hebdomadaires des Séances de l'Académie des Sciences*, **166**, 1918, 601.

[1882] Jordan, Camille. *Cours d'analyse de l'École Polytechnique*, first edition, 3 Volumes, Gauthier-Villars, 1882–1887.

[1893] Jordan, Camille. *Cours d'analyse de l'École Polytechnique*, second edition, 3 Volumes, Gauthier-Villars, 1893–1896.

[1913] Julia, Gaston. "Sur les lignes singulières des fonctions analytiques." *Bulletin de la Société mathematique de France*, **41**, 1913, 351–366.

[1917] Julia, Gaston. "Sur les substitutions rationnelles." *Comptes rendus hebdomadaires des Séances de l'Académie des Sciences*, **166**, 599-601.

[1918] Julia, Gaston. " Mémoire sur l'iteration des fonctions rationnelles." *Journal dc Mathématiques pures et appliquées*, **8**, 1918, 47–245. Also in [1968,I:121–319].

[1918a] Julia, Gaston. "Sur l'itération des fractions rationnelles." *Comptes rendus hebdomadaires des Séances de l'Académie des Sciences*, **166**, 1918, 61–64.

[1918b] Julia, Gaston. "Sur des problèmes concernant l'itération des fractions rationnelles." *Comptes rendus hebdomadaires des Séances de l'Académie des Sciences*, **166**, 1918, 153–166.

[1918c] Julia, Gaston. "Sur les substitutions rationnelles." *Comptes rendus hebdomadaires des Séances de l'Académie des Sciences*, **166**, 1918, 599-601.

[1919a] Julia, Gaston. "Sur quelques problèms relatifs à l'itération des fractions rationnells." *Comptes rendus hebdomadaires des Séances de l'Académie des Sciences*, **168**, 1919, 147–49.

[1919b] Julia, Gaston. "Une propriété générale des fonctions entières liées au thérème de M. Picard." *Comptes rendus hebdomadaires des Séances de l'Académie des Sciences*, **168**, 1919, 502.

[1919c] Julia, Gaston. "Quelques propriétés générales des fonctiones entières liées au thérème de M. Picard." *Comptes rendus hebdomadaires des Séances de l'Académie des Sciences*, **168**, 1919, 598.

[1919d] Julia, Gaston. "Mémoire relatif à l'étude des substitutions rationnelles à une variable." *Bulletin de la Société mathematique de France*, Series 2, **43**, 1919, 106–09.

[1920a] Julia, Gaston. "Sur les familles de fonctions de plusieurs variables." *Comptes rendus hebdomadaires des Séances de l'Académie des Sciences*, **170**, 1920, 875.

[1920c] Julia, Gaston. "Sur les familles de fonctions de plusieurs variables." *Comptes rendus hebdomadaires des Séances de l'Académie des Sciences*, **170**, 1920, 1040.

[1920d] Julia, Gaston. "Sur les familles de fonctions de plusieurs variables." *Comptes rendus hebdomadaires des Séances de l'Académie des Sciences*, **170**, 1920, 1234.

[1921a] Julia, Gaston. "Sur une classe d'équation fonctionnelle." *Comptes rendus hebdomadaires des Séances de l'Académie des Sciences*, **173**, 1921, 813–16.

[1921b] Julia, Gaston. "Sur les entières ou méromorphes." *Comptes rendus hebdomadaires des Séances de l'Académie des Sciences*, **173**, 1921, 964–67.

[1921c] Julia, Gaston. "Sur les solutions méromorphes de certaines équations fonctionnelles." *Comptes rendus hebdomadaires des Séances de l'Académie des Sciences*, **173**, 1921, 1149-52.

[1922a] Julia, Gaston. "Les équations functionnelles et la represéntation conforme." *Comptes rendus hebdomadaires des Séances de l'Académie des Sciences*, **174**, 1922, 517–19.

[1922b] Julia, Gaston. "Nouvelles applications de la représentation conform aux équations fonctionnelles." *Comptes rendus hebdomadaires des Séances de l'Académie des Sciences*, **174**, 1922, 653–655.

[1922c] Julia, Gaston. "Sur la tranformation de substitutions rationnelles en substitutions linéaires." *Comptes rendus hebdomadaires des Séances de l'Académie des Sciences*, **174**, 1922, 800–02.

[1922d] Julia, Gaston. "Sur une équation aux dérivées fonctionnelles liée a représentation conforme." *Annales Scientifiques de l'École Normale Supérieure*, **39**, 1922, 1–28.

[1922e] Julia, Gaston. "Mémoire sur la permutabilité des fractions rationnelles." *Annales Scientifiques de l'École Normale Supérieure*, **39**, 1922, 131–215.

[1922f] Julia, Gaston. "Sur les substitutions rationnelles à deux variable." *Comptes rendus hebdomadaires des Séances de l'Académie des Sciences*, **174**, 1922, 1182–1185.

[1923a] Julia, Gaston. "Sur les substitutions rationnelles à deux variables." *Comptes rendus hebdomadaires des Séances de l'Académie des Sciences*, **175**, 1923, 58.

[1923b] Julia, Gaston. "Sur une classe d'equations fonctionnelles." *Annales Scientifiques de l'École Normale Supérieure*, **40**, 1923, 97–150.

[1934] Julia, Gaston. *Notice sur les travaux Scientifiques.* Also in [1968,I:1–88].

[1968] Julia, Gaston. *Oeuvres de Gaston Julia*, Gauthier-Villars, Paris, 1968–1970.

[1913] Kasner, Edward. "Conformal Geometry." *Proceedings, International Congress of Mathematicians*, Volume II, Cambridge University Press, 1913.

[1992] Kellerstrom, Nick. "Thomas Simpson and 'Newton's method of approximation': an enduring myth." *British Journal of the History of Science*, **25**, 347–54.

[1972] Kline, Morris. *Mathematical Thought from Ancient to Modern Times*, Oxford University Press, New York, 1972.

[1906] Koch, Helge von. "Une méthod géométrique élémentaire pour l'étude de certaines questions de la théorie des courbes planes." *Acta Mathematica*, **30**, 1906, 145–76.

[1883] Koenigs, Gabriel. "Recherches sur le substitutions uniforms." *Bulletin des Sciences mathématiques et astronomiques*, Series 2, **7**, Part 1, 1883, 340–57.

[1884] Koenigs, Gabriel. "Recherches sur les inegrales de certaines èquations fonctionelles." *Annales Scientifiques de l'École Normale Supérieure*, Series 3, **1**, 1884, s1–s41.

[1884a] Koenigs, Gabriel. "Sur les intégrales de certaines équations fonctionnelles." *Comptes rendus hebdomadaires des Séances de l'Académie des Sciences*, **99**, 1884, 1016–17.

[1885] Koenigs, Gabriel. "Nouvelles recherches sur le équations fonctionelles." *Annales Scientifiques de l'École Normale Supérieure*, Series 3, **2**, 1885, 385–404.

[1885a] Koenigs, Gabriel. "Sur les conditions d'holomorphisme des intégrales de l'équation itérative, et de quelques autres équations fonctionnelles." *Comptes rendus hebdomadaires des Séances de l'Académie des Sciences*, **101**, 1885, 1137–39.

[1897] Koenigs, Gabriel. *Notice sur les trauvaux scientifiques de Gabriel Koenigs*, Deslis Frères, Imprimeurs, 1897.

[1971] Koppelman, Elaine. "The Calculus of Operations and the Rise of Abstract Algebra." *Archive for History of the Exact Sciences*, **8**, 1971, 155–242.

[1882] Korkine, Alexándr. "Sur un problèm d'interpolation." *Bulletin des Sciences mathématiques et astronomiques*, Series 2, **6**, 1882, 228–242.

[1990] Kuczma, Marek and Choczewski, Bogdan and Ger, Roman. *Iterative Functional Equations, Volume 32, Encylopedia of Mathematics and its Applications*, Cambridge University Press, 1990.

[1985] Landau, Edmund. *Edmund Landau Collected Works*, Thomas Verlag, 1985.

[1904] Landau, Edmund. "Über eine Verallgemeinerung des Picard'schen Satzes." *Berliner Sitzungsberichte*, **38**, 1904, 1118–33. Also in [1985,II:129-44].

[1904] Landau, Edmund. "Über den Picard'schen Satz." *Vierteljahrschrift der Naturforschenden Gesellschaft in Zürich*, **51**, 1906, 252–318. Also in [1985,III:113–80].

[1907] Lattès, Samuel. "Sur les equations functionnelles qui définissent une courbe ou une surface invariante par une transformation." *Annali di matematica pure and applicatat*, **13**, 1907, 1–137.

[1908] Lattès, Samuel. "Nouvelle recherches sur les courbes invariant par une transformation $(X, Y; x, y, y')$." *Annales Scientifiques de l'École Normale Supérieure*, **25**, 1908, 221–55.

[1918a] Lattès, Samuel. "Sur l'itération des substitutions rationnelles et les fonctions de Poincaré." *Comptes rendus hebdomadaires des Séances de l'Académie des Sciences*, **166**, 1918, 26–28.

[1918b] Lattès, Samuel. "Sur l'itération des substitutions rationelles à deux varibles." *Comptes rendus hebdomadaires des Séances de l'Académie des Sciences*, **166**, 1918, 151–53.

[1918c] Lattès, Samuel. "Sur l'itération des fractions rationnelles." *Comptes rendus hebdomadaires des Séances de l'Académie des Sciences*, **166**, 1918, 486–89.

[1931] Launay, Louis. Necrology of Gabriel Koenigs. *Compte rendus, l'Académie des Sciences*, **193**, 1931, 57–58.

[1895] Leau, Léopold. "Sur les équations fonctionnelles." *Comptes rendus hebdomadaires des Séances de l'Académie des Sciences*, **120**, 1895, 427–29.

[1897] Leau, Léopold. *Étude sur les équations fonctionnelles a une ou a plusieürs variables.* Gauthier-Villars, 1897.

[1898] Leau, Léopold. "Sur un probléme d'itération." *Bulletin de la Société mathématique de France*, **26**, 5–9, 1898.

[1902] Lebesgue, Henri. "Intégrale, longuer, aire." *Annali di Matematica pura ed applicata*, Series 3, **7**, 1902, 201–331.

[1922] Lebesgue, Henri. *Notice sur les travaux scientifiques de M. Henri Lebesgue*, Toulouse, 1922.

[1895] Lémeray, Ernest. "Un théorème sur les fonctions itératives." *Bulletin de la Société mathématique de France*, **23**, 255–63, 1895.

[1896a] Lémeray, Ernest. "Sur la dérivée des fonctions interpolées." *Nouvelles annales de Mathématiques*, **15**, 325–27, 1896.

[1896b] Lémeray, Ernest. "Sur la convergence des substitutions uniforms." *Comptes rendus hebdomadaires des Séances de l'Académie des Sciences*, **123**, 793–94, 1896.

[1897a] Lémeray, Ernest. "Sur la dérivée des fonctions itératives au point limite." *Bulletin de la Société mathématique de France*, **25**, 51–53, 1897.

[1897b] Lémeray, Ernest. "Dérivée des fonctions itératives par rapport à l'indice d'itération." *Bulletin de la Société mathématique de France*, **25**, 92–94, 1897.

[1897c] Lémeray, Ernest. "Sur la convergence des substitutions uniforms." *Nouvelles annales de Mathématiques*, **16**, 306–19, 1897.

[1897d] Lémeray, Ernest. "Racines de quelques équations transendantes." *Nouvelles annales de Mathématiques*, **16**, 540–46, 1897.

[1897e] Lémeray, Ernest. "Sur les convergence des substitutions uniforms." *Comptes rendus hebdomadaires des Séances de l'Académie des Sciences*, **124**, 1220–22, 1897.

[1897f] Lémeray, Ernest. "Sur les équations fonctionelles linéaires." *Comptes rendus hebdomadaires des Séances de l'Académie des Sciences*, **125**, 1160–61, 1897.

[1898a] Lémeray, Ernest. "Sur la convergence des substitutions uniformes." *Nouvelles annales de Mathématiques*, **17**, 75–80, 1898.

[1898b] Lémeray, Ernest. "Sur le calcul des racines des équations par approximations successives." *Nouvelles annales de Mathématiques*, Series 3, **17**, 1898, 534-39.

[1898c] Lémeray, Ernest. "Sur quelques algorithmes généraux et sur l'itération." *Bulletin de la Société mathématique de France*, **26**, 1898, 10–15.

[1898d] Lémeray, Ernest. "Sur quelques algorithmes généraux et sur l'itération." *Comptes rendus hebdomadaires des Séances de l'Académie des Sciences*, **126**, 1898, 510–12.

[1898e] Lémeray, Ernest. "Sur certaines équations fonctionnelles linéaires." *Comptes rendus hebdomadaires des Séances de l'Académie des Sciences*, **126**, 1898, 949-50.

[1899a] Lémeray, Ernest. "Sur les équations fonctionnelles qui caractérisent les opérations associatives et les opérations distributives." *Bulletin de la Société mathématique de France*, **27**, 1899, 130–3.

[1899b] Lémeray, Ernest. "Application des fonctions doublement périodiques a la solution d'un probléme d'itération." *Bulletin de la Société mathématique de France*, **27**, 1899, 282–84.

[1912] Lévy, Paul. "Remarques sur le théorème de M. Picard." *Bulletin de la Société mathématique de France*, **40**, 1912, 26–39.

[1877] Lie, Sophus. Untitled summary of his "Théorie des groupes transformation." *Bulletin des Sciences mathématiques et astronomiques*, Series 2, **1**, Part 2, 382–83.

[1908] Lindelöf, Ernst. "Mémoire sur certaines inégalités dans la théorie des fonctions monogènes." *Acta Societatis Scientiarium Fennicæ*, **35**, 1908, 3–35.

[1910] Lindelöf, Ernst. "Sur le théorème de M. Picard dans la théorie des fonctions mongènes." *Compte rendu du Congrès des Mathematiciens, Stockholm*, 1910, 112–136.

[1900] Lorch, Edgar R. "Joseph Fels Ritt." *Bulletin of the American Mathematical Society*, **57**, 1951, 307–318.

[1980] Lorch, Edgar R. "Joseph Fels Ritt" from *The Dictionary of Scientific Biography*, Charles Schribners and Sons, Volume 11, 1980, 470–71.

[1973] Mandelbrojt, Szolem. "Notice nécrologique sur Maurice Fréchet." *Comptes rendus hebdomadaires des Séances de l'Académie des Sciences, Vie Académique*, **277**, 1973, 73–76.

[1974] Mandelbrojt, Szolem. "Notice nécrologique sur Paul Montel." *Comptes rendus hebdomadaires des Séances de l'Académie des Sciences, Vie Académique*, **280**, 1974, 186–88.

[1990] Milnor, John. *Dynamics in One Complex Variable: Introductory Lectures.* SUNY Stony Brook Institute for Mathematical Sciences, 1990, preprint.

[1975] Monna, A. F. *Dirichlet's Principle*, Oosthoeek, Scheltema and Wolkema, Utrecht, 1975.

[1903] Montel, Paul. "Sur íntegrabilitie d'une expresion différentielle." *Comptes rendus hebdomadaires des Séances de l'Académie des Sciences*, **136**, 1903, 1233–35.

[1904] Montel, Paul. "Sur les suites de fonctions analytiques." *Comptes rendus hebdomadaires des Séances de l'Académie des Sciences*, **138**, 1904, 469-71.

[1906] Montel, Paul. "Sur les séries de fonctions analytiques." *Bulletin des Sciences mathematiques et astronomiques*, **30**, 1906, 189-92.

[1907a] Montel, Paul. "Sur les points irréguliers des séries convergentes de fonctions analytiques." *Comptes rendus hebdomadaires des Séances de l'Académie des Sciences*, **145**, 1907, 910–13.

[1907b] Montel, Paul. "Sur les suites infinies des fonctions." Gauthier-Villars, Paris, 1907.

[1910] Montel, Paul. *Leçons sur les séries de polynomes a une variable complexe*, Gauthier-Villars, Paris, 1910.

[1911a] Montel, Paul. "Sur les fonctions analytiques admettent deux valeurs exceptionnelles dans un domain." *Comptes rendus hebdomadaires des Séances de l'Académie des Sciences*, **153**, 1911, 996–98.

[1911b] Montel, Paul. "Sur l'indétermination d'une fonction uniforme dans le voisinage de ses points essentiels." *Comptes rendus hebdomadaires des Séances de l'Académie des Sciences*, **153**, 1911, 1455–56.

[1912a] Montel, Paul. "Sur les familles de fonctions analytiques qui addmettent des valeurs exceptionnelles dans un domaine." *Annales Scientifiques de l'École Normale Supérieure*, **29**, 1912, 487–535.

[1912b] Montel, Paul. "Sur quelques généralisations des théorèmes de M. Picard." *Comptes rendus hebdomadaires des Séances de l'Académie des Sciences*, **155**, 1912, 1000–03.

[1912c] Montel, Paul. "Sur l'existence des derivés." *Comptes rendus hebdomadaires des Séances de l'Académie des Sciences*, **155**, 1912, 1478–80.

[1916] Montel, Paul. "Sur les familles normales de fonctions analytique." *Annales Scientifiques de l'École Normale Supérieure*, **33**, 1916, 223–302.

[1917] Montel, Paul. "Sur la représentation conforme." *Journal de Mathématiques pures et appliquées*, Series 7, **3**, 1917, 1–54.

[1927] Montel, Paul. *Familles normale*, reprint, Chelsea Publishing Company, New York, 1974.

[1947] Montel, Paul. *Selecta cinquantenaire scientifique de M. Paul Montel*, Gauthier-Villars, Paris, 1947.

[1980] Nathan, Henry. "Pierre Fatou" from *The Dictionary of Scientific Biography*, Charles Schribners and Sons, Volume 4, 1980, 547–48.

[1887a] Netto, Eugen. "Ueber einen Algorithmus zur Auflösung numerischer algebraischer Gleichungen." *Mathematische Annalen*, **29**, 1887, 140–147.

[1887b] Netto, Eugen. "Zur Theoreie der iterirten Functionen." *Mathematische Annalen*, **29**, 1887, 148–153.

[1929] Nevanlinna, Rolf. *Le théorème de Picard-Borel et la théorie des fonctions méromorphes*, reprint, Chelsea Publishing Company, New York, 1974.

[1927] Ortvay, R. Title Unknown. *Mathematikai es phys lapok.* **34**, 1927, 5–25.

[1902] Osgood, William. "Notes on the Functions Defined by an Infinite Series Whose Terms are Analytic Functions of a Complex Variable." *Annals of Mathematics*, Series 2 **3**, 1902, 25ff.

[1968] Ozhigova, E. P. *Alexandr Nikolaevich Korkine*, Leningrad, 1968.

[1932] P. D. E. "Gabriel Koenigs." *Revista Mathemática Hispano-Americana*, Series 2, **7**, 1932, 85–86.

[1887] Peano, Giuseppe. *Applicazione geometriche del calcolo infinitesimale*, Bocca, Torino, 1887.

[1989] Peitgen, Heinz-Otto, ed. *Newton's Method and Dynamical Systems*, Klummer Academic Publishers, 1989.

[1984] H. Peitgen, D. Saupe, and F. v. Haessler. "Cayley's Problem and Julia Sets." *Math Intel*, **6**, 1984, 11–20.

[1986] Peitgen, Heinz-Otto and Richter, Peter. *The Beauty of Fractals.* Springer-Verlag, 1986.

[1917] Pfeiffer, George. "On the Conformal Mapping of Curvilinear Angles. The functional equation $\phi[f(x)] = a_1\phi(x)$." *Transactions of the American Mathematical Society*, **18**, 185–98.

[1978] Picard, Émile. *Oeuvres de Émile Picard, Volume 1*, Centre National de la Recherche Scientifique, Paris, 1978.

[1879a] Picard, Émile. "Sur une propriété des fonctions entières." *Comptes rendus hebdomadaires des Séances de l'Académie des Sciences*, **88**, 1879, 1024–27. Also in [1978:19-22].

[1879b] Picard, Émile. "Sur les fonctions entières." *Comptes rendus hebdomadaires des Séances de l'Académie des Sciences*, **89**, 1879, 662–65. Also in [1978:23–25].

[1879c] Picard, Émile. "Sur les fonctions analytiques uniformes dans le voisinage d'un point singulier essentiel." *Comptes rendus hebdomadaires des Séances de l'Académie des Sciences*, **89**, 1879, 745–47. Also in [1978:27–29].

[1880] Picard, Émile. "M'emoire sur les fonctions entières." *Annales Scientifiques de l'École Normale Supérieure*, **9**, 1880, 145–66. Also in [1978:39-60].

[1900] Picard, Émile. "Sur une class de surfaces algbriques dont les coordonées s'expriment par les des fonctions uniformes de deux paramètres." *Annales Scientifiques de l'École Normale Supérieure*, **28**, 1900, 17–25. Also in [1978:201–09].

[1917] Picard, Émile. *Les sciences mathématiques en France depuis un demi-siècle*, Gauthier-Villars, Paris, 1917.

[1928] Picard, Émile. "Leçons sur quelques' équations fonctionelles avec des applications." *Cahiers scientifiques 3*, Paris, 1928.

[1906] Pincherle, Salvatore. "Funktionaloperationen und-Gleichungen" from *Encyklopädie der Mathematischen Wissenschaften*, Volume II, Leipzig, 1906, 761—817.

[1912] Pincherle, Salvatore. "Equations et opérations functionnelles" from *Encyclopédie des sciences mathématiques pures et appliquées*, Volume II, Leipzig, 1912.

[1917] Pincherle, Salvatore. "Sulle catene di radicali quadratici." *Atti della reale Accademi delle scienze di Torino*, **53**, 1917/1918, 745–63.

[1918a] Pincherle, Salvatore. "Sulle radici reali delle equazioni iterate di una equazione quadratica." *Rendiconti delle sedute della reale Accademia dei Lincei*, Series 5, **27**, 1918, 177–83.

[1918b] Pincherle, Salvatore. "Sull'iterazione della funzione $x^2 - -a$." *Rendiconti delle sedute della reale Accademia dei Lincei*, Series 5, **27**, 337–43.

[1920a] Pincherle, Salvatore. "L'iterazione completa di $x^2 - -a$." *Rendiconti delle sedute della reale Accademia dei Lincei*, Series 5, **29**, 1920, 329-33.

[1920b] Pincherle, Salvatore. "Sulla funzione iterata di una razionale interra." *Rendiconti delle sedute della reale Accademia dei Lincei*, Series 5, **29**, 1920, 403–407.

[1920c] Pincherle, Salvatore. "Sopra alcune equazioni funzionali." *Rendiconti delle sedute della reale Accademia dei Lincei*, Series 5, **29**, 1920, 279-81.

[1922] Poggendorff, J.C. *Biographisch-literarisches Handwortenbuch fur Matematik, Asronomie, Physik, Chemie und verwandte Wissenschaftengebiete (1906–1922)*, Leipzig, 1922.

[1881] Poincaré, Henri. "Memoire sur les courbes définies par une équation différentielle." *Journal de Mathématiques pures et appliquées*, Series 3, **7**, 1881, 375–422.

[1890] Poincaré, Henri. "Sur une class nouvelle de transcendantes uniformes." *Journal de Mathématiques pures et appliquées*, Series 4, **6**, 1890, 313–65.

[1892] Poincaré, Henri. *Les méthodes nouvelles de la méchanique céleste*, 3 volumes, Gauthier-Villars, Paris, 1892, 1893, 1898.

[1907] Posse, K. Title unknown. *Soobcheniya (i protokol) Kharkov University*, Series 2, **10**, 1907, 217–30.

[1908] Posse, K. Title Unknown. *Zh minist norodnykh proseshcheniya*, Saint Petersburg **11**, 1908, 25–46.

[1909] Posse, K. Title Unknown. *Mathematicheski Sbornski*, Series 1, 1909, 1–27.

[1690] Raphson, Joseph. *Analysis Aequationum Universalis*, 1690.

[1881] Rausenberger, Otto. "Theorie der allgemeinen Periodizität." *Mathematische Annalen*, **18**, 1881, 379-410.

[1917] Ritt, Joseph Fels."On Certain Real Solutions of Babbage's Functional Equation." *Annals of Mathematics*, Series 2, **17**, 1915, 113–123.

[1918] Ritt, Joseph Fels. "Sur l'itération des fonctionnelles rationnelles." *Comptes rendus hebdomadaires des Séances de l'Académie des Sciences*, **166**, 1918, 380–81.

[1920] Ritt, Joseph Fels. "On the Iteration of Rational Functions." *Transactions of the American Mathematical Society*, **21**, 1920, 348–56.

[1922] Ritt, Joseph Fels. "Periodic Functions with a Multiplication Theorem." *Tranactions of the American Mathematical Society*, **23**, 16–25.

[1923] Ritt, Joseph Fels. "Sur les fonctions rationnelles permutables." *Comptes rendus hebdomadaires des Séances de l'Académie des Sciences*, **176**, 60–1, 1923.

[1904] Schottky, Friedrich. "Über den Picard'schen Satz und die Borel'schen Ungleichungen." *Sitzungsberichte der Königlich Preussischen Akademie der Wissenschaften*, **42**, 1904, 1244–62.

[1907] Schottky, Friedrich. "Über zwei Beweise des allgemeinen Picard'schen Satzes." *Sitzungsberichte der Königlich Preussischen Akademie der Wissenschaften*, **46**, 1907, 823–40.

[1862] Schröder, Ernst. "Über die Vielecke vov Gebrochener Seitenzahl oder die Bedeutung der Stern-Ploygone in der Geometrie." *Zeitschrift für Mathematik und Physik*, **7**, 1862, 55–64.

[1870] Schröder, Ernst. "Ueber unendlich viele Algorithmen zur Auflösung der Gleichungen." *Mathematische Annalen*, **2**, 1870, 317–65.

[1871] Schröder, Ernst. "Ueber iterite Functionen." *Mathematische Annalen*, **3**, 1871, 296–322.

[1981] Segal, Sanford. *Nine Introductions to Complex Analysis*, North-Holland Publishing Company, New York, 1981.

[1866/85] Serret, Joseph Alfred. *Cours d'algébre Supérieure, fifth edition*, 2 Volumes, Gauthier-Villars, Paris, 1885.

[1987] Shishikura, Mitsuhiro. "On the Quasiconformal Surgery of Rational Functions." *Annales Scientifiques de l'École Normale Supérieure*, **20**, 1987, 1–20.

[1942] Siegel, Carl. "Iteration of Analytic Functions." *Annals of Mathematics*, **43**, 1942, 607–12.

[1985] Smale, Stephen. "On the Efficiency of Algorithms of Analysis." *Bulletin of the American Mathematical Society*, **13**, 1985, 87–121.

[1875] Smith, Henry J. Stephen. "On the Integration of Discontinuous Functions." *Proceedings of the London Mathematical Society*, **6**, 1875, 140–53.

[1906] Spiess, O. "Theorie der linearen Iteralgleichung mit konstanten Koeffizienten." *Mathematische Annalen*, **62**, 1906, 226–52.

[1894] Stieltjes, Thomas-Jan. "Recherches sur les fractions continues." *Annales de la Faculté des Sciences de Toulouse*, **8**, 1894, J.1–J.122.

[1983] Sullivan, Dennis. "Conformal Dynamical Systems" from *Geometric Dynamics*, Springer-Verlag Lecture Notes, 1985, 725–52.

[1985] Sullivan, Dennis. "Quasiconformal Homeomorphisms and Dynamics I: Solution of the Fatou-Julia Problem on Wandering Domains." *Annals of Mathematics*, **122**, 1985, 401–18.

[1981] Targonski, Györrgy. *Topics in Iteration Theory*, Vandenhoeck and Ruprecht, Göttingen, 1981.

[1980] Taton, Rene. "Gabriel Koenigs" from *The Dictionary of Scientific Biography*, Charles Schribners and Sons, Volume 7, 1980, 446.

[1881] Voltera, Vito. "Alcune osservazioni sulle funzioni punteggiate discontinue." *Giornale di Matematiche*, **19**, 1881, 332–72.

[1895] Weierstrass, Karl. *Mathematische Werke*, Mayer and Müller, Berlin, 1895.

[1841] Weierstrass, Karl. "Zur Theorie der Potenzreihen." Unpublished, 1841. Also in [1895,I:67–71].

[1880] Weierstrass, Karl. "Zur Functionenlehre." *Monatsbericht der Königlich Akademie der Wissenschaften*, 1880. Also in [1895,II,201–30].

[1980] Wussing, Hans. "Ernst Schröder" from *The Dictionary of Scientific Biography*, Charles Schribners and Sons, Volume 12, 1980, 216–17.

[1980] Youschkevitch, A. P. "Pafnuty Chebyshev" from *The Dictionary of Scientific Biography*, Charles Schribners and Sons, Volume 3, 1980.

[1993] Ypma, Tjalling. "The Historical Development of the Newton-Raphson-Simpson Method." Preprint.

[1911] Zoretti, Ludovic. *Leçons sur le prolongment analytique*, Gauthier-Villars, 1911.

[1912] Zoretti, Ludovic. "Sur les ensembles de points," from *Encyclopédie des sciences mathématiques pures et appliquées*, Book II, Volume 1, Leipzig, 1912.

Index

Aspekte der Mathematik

Edited by Klas Diederich

Band D 1: H. Kraft: Geometrische Methoden in der Invariantentheorie

Band D 2: J. Bingener: Lokale Modulräume in der analytischen Geometrie 1

Band D 3: J. Bingener: Lokale Modulräume in der analytischen Geometrie 2

Band D 4: G. Barthel/F. Hirzebruch/T. Höfer: Geradenkonfigurationen und Algebraische Flächen*

Band D 5: H. Stieber: Existenz semiuniverseller Deformationen in der komplexen Analysis

Band D 6: I. Kersten: Brauergruppen von Körpern

*A Publication of the Max-Planck-Institut für Mathematik, Bonn

MIX
Papier aus verantwortungsvollen Quellen
Paper from responsible sources
FSC® C105338

If you have any concerns about our products,
you can contact us on
ProductSafety@springernature.com

In case Publisher is established outside the EU,
the EU authorized representative is:
Springer Nature Customer Service Center GmbH
Europaplatz 3, 69115 Heidelberg, Germany

Printed by Libri Plureos GmbH
in Hamburg, Germany